Basics of Horticulture

Basics of Horticulture

Sharon Pastor Simson
Martha C. Straus

Oxford Book Company
Jaipur ▪ New Delhi

Published By

Oxford Book Company

Regd. Off. "Bony Residency", Gate No. 2
Opp. Tilak Public School, Vishweswariya Nagar,
Gopalpura Road, Jaipur - 302018, India
Ph: 0141 - 2761280, Telefax: 2761381
email - oxfordbookjpr@gmail.com
website - oxfordbookcompany.com

Delhi Office: 102, 1st Floor, Satyam House,
4327/3, Ansari Road, Darya Ganj,
New Delhi - 110002, India
Ph: 011-45652440

Distributed By

KEYA INTERNATIONAL PRESS LLP
E-mail: keyainternationalpressllp@gmail.com

Published in India, 2025

ISBN : 9789380179186

Typeset by Shivangi Computers, Jaipur

Printed and bound in India by Thomson Press Ltd.,
New Delhi

Contents

Preface

India with diverse soil and climate comprising several agro-ecological regions provides ample opportunity to grow a variety of horticulture crops. These crops form a significant part of total agricultural produce in the country comprising of fruits, vegetables, root and tuber crops, flowers, ornamental plants, medicinal and aromatic plants, spices, condiments, plantation crops and mushrooms. It is estimated that all the horticulture crops put together cover nearly 11-6 million hectares area with an annual production of 91 million tonnes. Though these crops occupy hardly 7 per cent of the cropped area they contribute over 18 per cent to the gross agricultural output in the country.

Horticultural crops play a unique role in India's economy by improving the income of the rural people. Cultivation of these crops is labour intensive and as such they generate lot of employment opportunities for the rural population. Fruits and vegetables are also rich source of vitamins, minerals, proteins, carbohydrates etc. which are essential in human nutrition. Hence, these are referred to as protective foods and assumed great importance as nutritional security of the people. Thus, cultivation of horticultural crops plays a vital role in the prosperity of a nation and is directly linked with the health and happiness of the people.

The present book attempts to provide all fundamental aspects of horticulture. After the introductory chapter, it discusses systematically the topics like anatomy and morphology of crops, plant nutrition, soil fertility, methods of plant propagation, various horticultural practices, etc. Not only that, the state of the art developments in areas like establishment and management of gardens; weed, pest and disease management, harvesting and packaging of horticultural products; horticultural marketing trends, etc have also been reviewed in the book. By reading this book, students will learn how and why plants grow and develop and sailed.

Sharon Pastor Simson
Martha C. Straus

1

Introduction to Horticulture

Horticulture is the industry and science of plant cultivation. Horticulturists work and conduct research in the disciplines of plant propagation and cultivation, crop production, plant breeding and genetic engineering, plant biochemistry, and plant physiology. The work particularly involves fruits, berries, nuts, vegetables, flowers, trees, shrubs, and turf. Horticulturists work to improve crop yield, quality, nutritional value, and resistance to insects, diseases, and environmental stresses.

Horticulture involves eight areas of study, which can be grouped into two broad sections—ornamentals and edibles:

- Arboriculture the study and selection, planting, care, and removal of individual trees, shrubs, vines, and other perennial woody plants.
- Floriculture (production and marketing of floral crops).
- Landscape horticulture (production, marketing and maintenance of landscape plants).
- Olericulture (production and marketing of vegetables).
- Pomology (production and marketing of fruits).
- Viticulture (production and marketing of grapes).
- Postharvest physiology (involves maintaining quality and preventing spoilage of horticultural crops).

Horticulturists can work in industry, government or educational institutions or private collections. They can be cropping systems engineers, wholesale or retail business managers, propagators and tissue culture specialists (fruits, vegetables, ornamentals, and turf), crop inspectors, crop production advisers, extension specialists, plant breeders, research scientists, and of course, teachers. Disciplines which complement horticulture include biology, botany, entomology, chemistry, mathematics, genetics, physiology, statistics, computer science, and communications, garden design, planting design. Plant science and horticulture courses include: plant materials, plant propagation, tissue culture, crop production, post-harvest handling, plant breeding, pollination management, crop nutrition, entomology, plant pathology, economics, and business.

Horticulture is practiced in many gardens, "plant growth centres" and nurseries. Activities in nurseries range from preparing seeds and cuttings to growing fully mature plants. These are often sold or transferred to ornamental gardens or market gardens.

HISTORY OF HORTICULTURE

The origins of horticulture lie in the transition of human communities from nomadic hunter-gatherers to sedentary or semi-sedentary horticultural communities, cultivating a variety of crops on a small scale around their dwellings or in specialised plots visited occasionally during migrations from one area to the next. In forest areas such horticulture is often carried out in swiddens. A characteristic of horticultural communities is that useful trees are often to be found planted around communities or specially retained from the natural ecosystem.

Horticulture sometimes differs from agriculture in (1) a smaller scale of cultivation, using small plots of mixed crops rather than large field of single crops (2) the cultivation of a wider variety of crops, often including fruit trees. In pre-contact North America the semi-sedentary horticultural communities of the Eastern Woodlands (growing maize, squash and sunflower) contrasted markedly with the mobile hunter-gatherer communities of the Plains people. In Central America, Maya horticulture involved augmentation of the forest with useful trees such as papaya, avocado, cacao, ceiba and sapodilla. In the cornfields, multiple crops were grown such as beans, squash, pumpkins and chilli peppers, in some cultures tended mainly or exclusively by women.

BASIC PRINCIPLES OF HORTICULTURE

Basic Needs

Plant growth is a very complicated process, but it is not necessary to understand all the intricate details in order to successfully manage the urban forest. However, understanding the basic requirements of tree growth will help. The primary environmental factors are:

- Soil is a reservoir for nutrients and moisture, and provides mechanical support.
- Sunshine furnishes heat and light.
- Air supplies carbon dioxide and oxygen.

Plants in the wild obtain these basic needs from a complex, but finely balanced biological system. Cultivation often disturbs the natural balance. Any factor in the plant's environment that becomes less than optimum will limit its growth.

Naturally, quantity is important, but quality also affects the vigor of trees. Air pollution, construction activity, and soil compaction place stresses on trees. Healthy trees can better with stand these stresses year after year. This part describes principles of tree growth and their relationship with the environment. Knowing these basics will help the people responsible for tree management make wise decisions.

Relationship of Air and Water in the Soil

Soil provides a foundation for tree growth-structurally and biologically. Soil supports a tree's physical weight and resists the forces of wind. Soil al supplies water, air, and nutrients. The value of the soil as structural and biological support is related to soil structure. Soil structure influences plant growth, because it affects moisture, aeration, heat transfer, and the mechanical resistance to root growth. Most people think of soil as being a mixture of solid elements, such as minerals and organic matter. But, open spaces in soil, called pores, are equally important.

The size and distribution of soil pores affect the movement and availability of moisture and air through the soil. The air-water relationship is a critical one for optimum plant growth. The number of pore spaces must be adequate and the variety of sizes appropriate. Pore size determines the relative amounts of air and water that soil will hold. Gravity quickly drains water from larger pores, making them good sources of air. Smaller pores tend to hold water against the pull of gravity, making them good for water storage. Because plants need both air and water in the soil, a good balance of large and small pore spaces is required for optimum growth.

Sand, with its open structure of many large pores, is called a "coarse" soil. Sandy soils generally drain well. They contain ample amounts of air, but little water. "Fine" clay soils drain poorly and tend to hold water in their many small pore spaces. But, pores in clay soils can be so small and hold water so tightly that tiny feeder roots cannot extract the moisture. Organic matter in the soil affects both the physical and chemical properties of the soil. A fertile soil rich in organic matter is literally alive. Although insects and earthworms are the most obvious inhabitants, microorganisms, such as bacteria and fungi, constitute the largest population by weight in the soil community.

Organic matter in the soil comes from decomposed plant and animal tissue as well as the micro-organisms themselves. Organic matter enhances the aeration and water-holding capacity of the soil. It also affects the soil's chemical properties by supplying necessary plant nutrients. The phrase "effective rooting depth" describes the portion of the soil where conditions are favorable for root growth-most often in the top three feet of the soil. Effective rooting depth may be limited by circumstances that restrict soil porosity or hinder plant growth. Four relatively common problems are surface crusting, high water tables, poor sub-surface drainage and claypans or hardpans.

Soil Compaction

Soil compaction is the major cause of death or decline of mature trees where efforts have been made to save them. It poses a very serious threat to good soil structure. Delicate soil pores are easily crushed, decreasing their capacity for water and air movement and hindering root growth. Wet soil is particularly vulnerable, because water lubricates soil particles and loosens binding agents. Small particles slip between the larger particles, filling the pore spaces. Loose soils will compact more than tight soils, and soils that have a broad range of particle sizes can be more severely compacted than more uniform soils.

Few soils can withstand traffic without becoming severely compacted. Compaction depends not only on the amount of pressure exerted, but also on the duration and frequency of exertion. The heel of a shoe exerts force per unit of area equal to that of heavy equipment. Pressure spreads with depth, so the compacting effects of pedestrians and vehicles usually occur to the top four inches of the soil.

Compaction is easier to prevent than to remedy. Avoid cultivating wet soils. Keep foot and vehicle traffic over the roots to a minimum. Where traffic is necessary, confine it to a few paths that stay well away from the driplines of trees. Light-weight vehicles with large, smooth, low pressure tires compact the soil less than heavy vehicles with smaller, heavily treaded, high pressure tires. A thick, coarse mulch spread on the soil surface where travel must occur helps to disperse the load.

The Root System

A healthy root system serves multiple functions for the tree. Roots collect water and minerals from the soil, store nutrients, and anchor the tree. Fibrous roots are primarily responsible for the absorption of water and minerals from the soil. They most often grow in the top six inches of the soil. Woody roots provide a framework for anchoring the tree in the ground. They support the weight of the trunk and branches, including leaves, snow and ice. And, wind pressure adds to the load.

Woody roots radiate out horizontally and vertically, decreasing in size as they get farther from the trunk. Woody roots seldom penetrate the soil deeper than three feet. Because they are relatively shallow, roots are vulnerable to damage from activities on the surface. It's important to be aware of roots, because they are not easily seen, and structural deficiencies often remain unnoticed until the top of the tree dies back or the entire tree falls over.

Root defects commonly occur in two forms (Fig. 1). When roots grow in a circle around the root mass it's called girdling. Roots may also be kinked into a characteristic "j-hook" growth pattern. Improper treatment in the nursery often causes these types of defects. To avoid problems, inspect the roots in the nursery before purchasing trees. Girdling roots may also occur after the tree is planted into the landscape. This happens when roots grow around the edge of the planting pit instead of out into the field soil. A lack of structural integrity in the root system may cause the planting to fail early in the tree's life or much later, but failure is sure to occur.

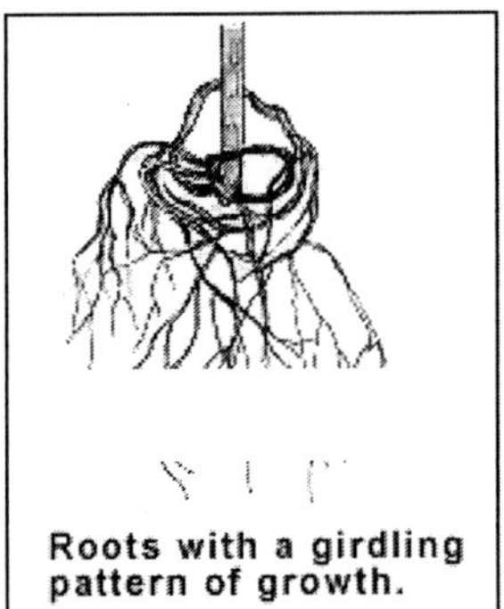

Roots with girdling pattern of growth

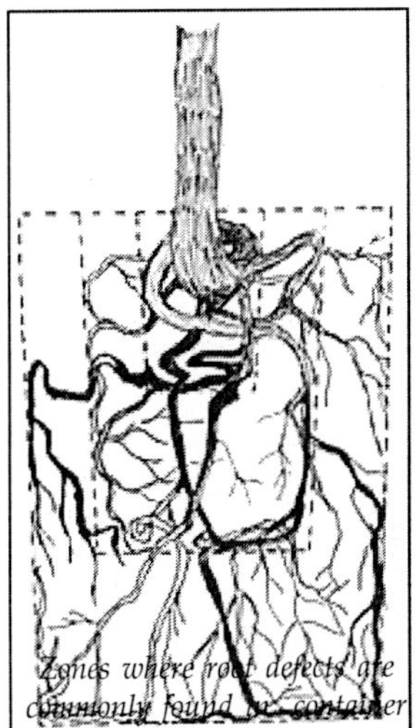

Zones where root defects are commonly found in container grown trees.

Figure 1. Types of root defects and the zones within the root mass where they may be found.

Structural Strength of Tree Crowns

A tree's crown must be strong enough to withstand wind forces, and carry the weight of snow, ice, and branches covered with thousands of leaves. To assess the structural strength of a crown, look at these characteristics: the spacing of lateral branches along the main trunk, the presence of a clearly dominant leader or main trunk, trunk taper near the ground, and angle of branch attachment. A tree with branches spread out along the trunk generally has better structure than one with all its branches growing close together on the trunk. A clearly dominant leader or main trunk provides a strong framework-called a scaffold-that supports lateral branches. Several leaders of similar size offer a weaker scaffold. Even round, oval, and vase shaped trees that commonly have more than one leader should show a clear hierarchy of leader sizes. This distinction of leader size is a sign of a good scaffold. Trunk taper at the base of the trunk helps support for the entire crown.

Young trees with no trunk taper usually have to be staked, so they don't topple over in high winds or under heavy loads of ice and snow. Finally, branches that have a wide angle of attachment or have connective tissue in the crotch are usually more structurally sound than those with sharp angles of attachment or imbedded bark in the crotch.

Plant Stress

Trees, regardless of species or location, have the same basic needs that must be met for the tree to remain healthy and grow. Basic requirements for plant growth include water, carbon dioxide, oxygen, nutrients, sunlight, appropriate temperature, and sufficient space. A balance must be maintained between the amount of water lost through the leaves and that taken up through the roots. Soil conditions must allow enough root growth to supply the leaves and branches with nutrients, gaseous exchange, and water. Photosynthesis must be able to supply the energy needs of the tree. And this energy must be conveyed from the leaves to the stem and roots. Last but not least, the strength with which the tree grows, known as vigor, must remain high enough to prevent attack by disease-causing agents, such as bacteria, fungi, and insects.

The urban environment often places tremendous stress on trees that natural defenses may not adequately protect against. The urban environment changes rapidly, at least in comparison to the life span of most trees. The amount of stress experienced by a tree is directly related to the rate of unfavorable change in its environment, the quality of the planting location, and the level of maintenance it receives.

Noninfectious diseases, which are not transmittable from a diseased plant to a healthy plant, will be especially threatening to trees whose defenses are weakened by urban pressures. Most urban stresses can be divided into the broad categories of environmental stress, animal injury, and people pressure. Most urban stresses can be divided into the broad categories of environmental stress, animal injury, and people pressure.

Environmental stress has the most severe effects on newly planted trees that have not yet adjusted to their new conditions and on established trees whose surroundings have changed, usually from the activities of people. Temperature stress results from extreme heat and cold as well as rapid changes. High light levels are usually associated with high temperatures. Thinning or clearing operations can expose remaining trees to excessive sunlight threatening them with light and temperature stress. Minimise the negative impacts of high temperature and light by planting shrubs and ground covers to create an "understory" beneath and around the trees. These new plants will help keep the area cooler. In areas surrounded by concrete or asphalt group the trees together or plant an understory to buffer high temperatures.

Low temperature can also injure trees. "Hardiness" refers to the adaptations of woody plants to cold conditions through patterns of growth and dormancy. While plants can often tolerate extreme cold during the dormant season, they could be injured by the same temperature during the growing season. Select plants with the extreme rather than the average temperature in mind. Plants in containers are more susceptible to extreme and rapidly changing temperatures, because their roots are not protected by the "thermal mass" of the earth.

Moisture stress is one of the most common problems. Too much or too little water both inhibit root function, and therefore, the growth of the entire plant. Physical damage from snow and ice, poor water quality, and erosion could also be classified as moisture stress. One of the most important factors contributing to moisture deficiency is the amount of space available for the roots. If roots are restricted, they cannot spread through enough soil to gather the moisture needed to support the crown.

Another type of moisture stress occurs mostly in broadleaved and needled evergreens. During the winter animals —both pets and wildlife and spring, the air may be windy and warm, but the soil frozen or dry. Known as "winter drying" conditions, more water is lost through leaves than is taken up by the roots.

Excess water from flooding, saturated soils or over watering can also place stress on the tree. Like other parts of the tree, roots need to ex change gases, such as oxygen and carbon dioxide, in order to survive. This exchange of gases occurs between the fibrous roots and the air in the soil's large pores. If the pores are filled with water, the roots suffocate. To avoid problems with excess moisture, determine the soil moisture and drainage patterns of the planting site,'select a tree that thrives under, or at least tolerates, those conditions, or modify the soil situation.

Soil stress can be classified as either chemical, biological, or physical. Chemical stress involves unfavorable pH, an imbalance in nutrients, or even the presence of toxic materials discharged into the soil by other plants or human recklessness. In cities, soil stress is usually related to physical changes affecting soil aeration and moisture. Physical stress involves something that restricts root growth, including streets, sidewalks, buildings, construction activities, construction debris in the soil, large boulders, trenching, grade changes, soil compaction, improperly used herbicides, and the roots of other trees.

Animals injure trees in several ways. Animal waste, especially urine, adds toxic compounds to the soil. Animal traffic compacts the soil. A host of animals —both pets and wildlife inflict direct injuries to the bark: birds peck it, cats shred it, and rodents chew it. Dogs are particularly destructive when chained to a tree. People apply a variety of pressures that adversely affect the health of trees. Trees are easily disturbed by construction of buildings and roads, soil compaction, chemicals, air pollution, and improper tree maintenance also fall into this category.

In addition to the stresses that come from the direct or indirect actions of people, urban trees also suffer from human inaction. Well meaning people sometimes impede appropriate urban forest management. Education is one way to increase the knowledge of urban forestry and to get people to work together to support it.

ENVIRONMENT IMPACTS ON HORTICULTURAL CROPS

Abiotic Factors

Weather and climate. Each horticultural crop has certain climatic requirements. Unfavourable weather and climatic conditions produce a stress. Components of weather and climate include:

Temperature

Temperature influences all physiological activities by controlling:

- Phytosynthesis
- Respiration
- Enzymic activity
- Organic matter decomposition
- Microbial growth and development
- Flowering
- Pollen viability
- Fruit set
- Hormonal balance
- Rate of maturation
- Rate of senescence
- Quality

- Yield
- Shelf life
- Harvest duration
- Vernalisation of biennials
- Seed germination
- Root development
- Water and nutrient absorption
- Pests and disease occurrence

Extremes of temperatures are:

- Frost damage at 0°C
- Chilling injury 0-2°C or lower but above freezing point.

Light

Photosynthesis uses light. Light intensity and duration are important for crop growth and development. Low light causes plants to be spindly, small leaves, bud blades, poor pollination and poor fruit quality. Plants differ in light requirements. An example was given for horticultural crops.

- Light affects pollen viability and fruit set.
- Certain seed require light to break dormancy e.g. lettuce.
- Some horticultural crops are sensitive to protoperiod.

Examples

- Short day plants flower rapidly when the days get shorter.
- Long day plants flower when days are longer.
- Short days hasten tuber formation in potatoes, root enlargement in sweet potatoes.
- Long days and high temperatures keep plants in staminate in Cucurbits.

Light intensity and duration are important for crop growth and development:

- Photosynthesis uses light
- Low light causes plants to be long and this (spindly), small leaves, bud blades, poor pollination and poor fruit quality.

- Photosynthesis is stopped at high light intensity depending on species.
- Plants differ in light requirements.
- Certain seeds require light and break dormancy.
- Some plants are sensitive to photoperiod or day length.
- Short day plants flower rapidly when the days get shorter.
- Long day plants flower when days are longer.
- Short days hasten tuber formation in potato, root enlargement in sweet potatoes.
- Long days and high temps keep plants in staminate in cucurbits.

Rainfall

- Water comprises more than 80% of the living plants, providing structural integrity for the plant.
- Rainfall natural source of water.
- Water is a major determinant of crop productivity and quality.
- Required in large quantities for plant growth than any other of the growth factor.
- Solvent for nutrients, minerals, etc..
- Water improves the germination of seeds.
- Essential to establish transplanted seedlings.
- Essential for faster establishment of the crop.
- Essential to facilitate application and distribution of fertilisers and pesticides.
- Provides protection in cold temp (frost).
- Facilitates harvesting of underground crops in dry soils.
- Cooling of the leaves during respiration.
- All crops have moisture range at which crop response is optimum.

Examples

- Crop differ in their tolerance of continuous wet conditions.
- Hydrophytes—aquatic.
- Mesophytes—most common of terrestrial plants.

- Xerophytes—can endure long period.
- Root systems affect the amount of water uptake..

Negative aspects of rainfall are:

- Waterlogging
- Drought
- Storms

Wind

- A slight wind is necessary to replenish Carbondioxide near plant surface.
- Wind carries oxygen away from the plant.
- Less wind, less evaporation, less water requirement.
- Wind can be a limiting factor in vegetables production where strong winds (a ver. Wind speed of 7.2 km/hr) occur e.g. Typhoons (wind speed of 60 km/hr)
- Use of windbreaks (shelter belts) minimise damage, a relatively slow wind.
- Dusting plants—affects rate of photosynthesis of coating the photosynthesizing leaves and green stem.
- May blow away pests and diseases thus carry pests to or away from crop.
- May interfere with farm operations e.g. spraying, etc..
- May aid in pollination.

Soils

Horticultural crops are adapted to a wide range of soil types at times with specific requirements. Soils particle types include clay 0.002 mm or less, silt 0.05 to 0.002 mm and sand 0.05 mm or above.

Biotic factors

- Insect pests
- Bacteria
- Fungi
- Nematodes

- Viruses
- Weeds
- Domestic and wild animals

Other Factors Internal to Plants

- Leaf structure
- Chlorophyll content
- Water conducting ability
- Osmotic adjustment
- Presence of strong sinks

Socio-economic Factors

- Poor road network
- Lack of credit facilities
- Poor marketing systems
- Poor education background
- Lack of availability of inputs-seeds, fertilisers, pesticides.
- Inadequate information on prices of inputs and produce.

Classification of Vegetable Crops

Classification based on edible part

a) Root vegetables (carrots, beetroot, radish)

b) Tuber vegetables (Yam, Irish potato)

c) Corm (Taro)

d) Stem vegetable (Asparagus)

e) Leaf vegetables (Lettuce, Amaranth)

f) Bulb vegetable (Onion, garlic)

g) Fruit vegetable (Tomato, okra)

h) Flower vegetables (Broccoli, Cauliflower)

i) Seed vegetable (Berus, Peas)

Classification based on botany

a) Solanaceae

b) Cucurbitaceae

Based on life cycle

a) Annuals (Amaranthus, Pumpkins)

b) Biennials (Cabbage, Beet)

c) Perennial (Chive, Asparagus)

Based on use

a) Greens

b) Salads

Based on climatic adaptation

a) Warm season crops (Okra, Pepper)

b) Cool season crops (Cauliflower, Broccoli)

Based on origin

a) Exotic vegetables (Cabbage, Tomato)

b) Indigenous vegetables (Amaranthus, Cleome)

The implications of each classification in relation to crop management.

VEGETABLE PRODUCTION SYSTEMS

To expose participants to the different vegetable production systems.

- Home garden, nutrition garden, community garden as source of food, nutrients.
- Market gardens
- Processing garden
- Truck gardening
- Export gardens
- Protected gardening
- Nursery gardening

- Seed production garden
- Vegetable forcing

The above factors were aimed at providing a perspective of the major practical applications of factors that would enhance production of horticultural crops. The objective was to develop an extended and in-depth understanding of the principal factors so as to be able to manipulate them to optimise vegetable production.

REFERENCES

Adams, C K, 2008. *Principles of Horticulture*. Butterworth-Heinemann.

McGee, J.R. and Kruse, M. 1986. "Swidden horticulture among the Lacandon Maya". University of California, Berkeley: Extension Media Center,.

Thompson, S.I. 1977. "Women, Horticulture, and Society in Tropical America". *American Anthropologist*, N.S., 79: 908-910

2

Plant Anatomy and Morphology

Horticulture (Lat. *hortus*, a garden), the art and science of the cultivation of garden plants, whether for utilitarian or for decorative purposes. The subject naturally divides itself into two sections, which we here propose to treat separately, commencing with the science, and passing on to the practice of the cultivation of flowers, fruits and vegetables as applicable to the home garden. The point of view taken is necessarily, as a rule, that of a British gardener.

Horticulture, apart from the mechanical details connected with the maintenance of a garden and its appurtenances, may be considered as the application of the principles of plant physiology to the cultivation of plants from all parts of the globe, and from various altitudes, soils and situations. The lessons derived from the abstract principles enunciated by the physiologist, the chemist and the physicist require, however, to be modified to suit the special circumstances of plants under cultivation.

The necessity for this modification arises from the fact that such plants are subjected to conditions more or less unnatural to them, and that they are grown for special purposes which are at variance, in degree at any rate, with their natural requirements. The life of the plant makes itself manifest in the processes of growth, development and reproduction. By growth is here meant mere increase in bulk, and by development the series of gradual modifications by which a plant, originally simple in its structure and conformation, becomes eventually complicated, and endowed with distinct parts or organs.

The reproduction of the higher plants takes place either asexually by the formation of buds or organs answering thereto, or sexually by the production of an embryo plant within the seed. The conditions requisite for the growth, development and reproduction of plants are, in general terms, exposure, at the proper time, to suitable amounts of light, heat and moisture, and a due supply of appropriate food.

The various amounts of these needed in different cases have to be adjusted by the gardener, according to the nature of the plant, its "habit" or general mode of growth in its native country, and the influence to which it is there subjected, as also in accordance with the purposes for which it is to be cultivated. It is but rarely that direct information on all these points can be obtained; but inference from previous experience, especially with regard to allied forms, will go far to supply such deficiencies. Moreover, it must be remembered that the conditions most favourable to plants are not always those to which they are subjected in nature, for, owing to the competition of other forms in the struggle for existence, liability to injury from insects, and other adverse circumstances, plants may actually be excluded from the localities best suited for their development.

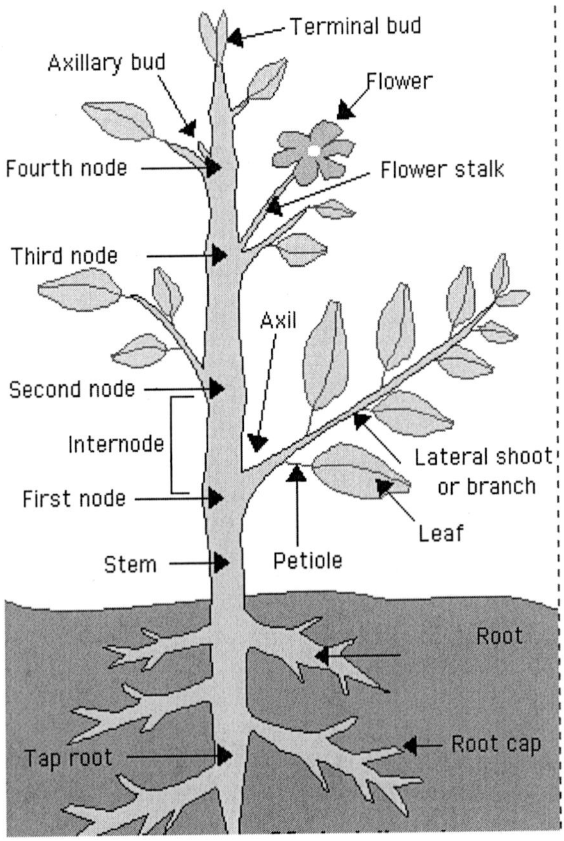

Figure 1. Structure of a plant

The gardener therefore may, and does, by modifying, improve upon the conditions under which a plant naturally exists. Thus it frequently happens that in our gardens flowers have a beauty and a fragrance, and fruits a size and savour denied to them in their native haunts. It behooves the judicious gardener, then, not to be too slavish in his attempts to imitate natural conditions, and to bear in mind that such attempts sometimes end in failure. The most successful gardening is that which turns to the best account the plastic organisation of the plant, and enables it to develop and multiply as perfectly as possible. Experience, coupled with observation and reflection, as well as the more indirect teachings of tradition, are therefore of primary importance to the practical gardener.

THE ROOT

The root, though not precluded from access of air, is not directly dependent for its growth on the agency of light. The efficiency of drainage, digging, hoeing and like operations is accounted for by the manner in which they promote aeration of the soil, raise its temperature and remove its stagnant water. Owing to their growth in length at, or rather in the immediate vicinity of, their tips, roots are enabled to traverse long distances by surmounting some obstacles, penetrating others, and insinuating themselves into narrow crevices. As they have no power of absorbing solid materials, their food must be of a liquid or gaseous character. It is taken up from the interstices between the particles of soil exclusively by the finest subdivisions of the fibrils, and in many cases by the extremely delicate thread-like cells which project from them and which are known as root-hairs.

The importance of the root-fibres, or "feeding roots" justifies the care which is taken by every good gardener to secure their fullest development, and to prevent as far as possible any injury to them in digging, potting and transplanting, such operations being therefore least prejudicial at seasons when the plant is in a state of comparative rest.

Root-Pruning and Lifting

In apparent disregard of the general rule just enunciated is the practice of root-pruning fruit trees, when, from the formation of wood being more active than that of fruit, they bear badly. The contrariety is more apparent than real, as the operation consists in the removal of the coarser roots, a process which results in the development of a mass of fine feeding roots. Moreover, there is a generally recognised quasi-antagonism between the vegetative and reproductive processes, so that, other things being equal, anything that checks the one helps forward the other.

Watering

So far as practical gardening is concerned, feeding by the roots after they have been placed in suitable soil is confined principally to the administration of water and, under certain circumstances, of liquid or chemical manure; and no operations demand more judicious management. The amount of water required, and the times when it should be applied, vary greatly according to the kind of plant and the object for which it is grown, the season, the supply of heat and light, and numerous other conditions, the influence of which is to be learnt by experience only. The same may be said with respect to the application of manures. The watering of pot-plants requires especial care. Water should as a rule be used at a temperature not lower than that of the surrounding atmosphere, and preferably after exposure for some time to the air.

Bottom-Heat

The "optimum" temperature, or that best suited to promote the general activity of roots, and indeed of all vegetable organs, necessarily varies very much with the nature of the plant, and the circumstances in which it is placed, and is ascertained by practical experience. Artificial heat applied to the roots, called by gardeners "bottom-heat," is supplied by fermenting materials such as stable manure, leaves, or by hot-water pipes. In winter the temperature of the soil, out of doors, beyond a certain depth is usually higher than that of the atmosphere, so that the roots are in a warmer and more uniform medium than are the upper parts of the plant. Often the escape of heat from the soil is prevented by "mulching," i.e. by depositing on it a layer of litter, straw, dead leaves and the like.

The Stem and its' subdivisions or branches raise to the light and air the leaves and flowers, serve as channels for the passage to them of fluids from the roots, and act as reservoirs for nutritive substances. Their functions in annual, biennial and herbaceous perennial plants cease after the ripening of the seed, whilst in plants of longer duration layer after layer of strong woody tissue is formed, which enables them to bear the strains which the weight of foliage and the exposure to wind entail.

The gardener aims usually at producing stout, robust, short-jointed stems, instead of long lanky growths defective in woody tissue. To secure these conditions free exposure to light and air is requisite; but in the case of coppices and woods, or where long straight spars are needed by the forester, plants are allowed to grow thickly so as to ensure development in an upward rather than in a lateral direction. This and like matters will, however, be more fitly considered in dealing hereafter with the buds and their treatment.

LEAVES

The work of the leaves may briefly be stated to consist of the processes of nutrition, respiration and transpiration. Nutrition *(assimilation)* by the leaves includes the inhalation of air, and the interaction under the influence of light and in the presence of chlorophyll of the carbon dioxide of the air with the water received from the root, to form carbonaceous food. Respiration in plants, as in other organisms, is a process that goes on by night as well as by day and consists in plants in the breaking up of the complex carbonaceous substances formed by assimilation into less complex and more transportable substances. This process, which is as yet imperfectly understood, is attended by the consumption of oxygen, the liberation of energy in the form of heat, and the exhalation of carbon dioxide and water vapour. Transpiration is loss of water by the plant by evaporation, chiefly from the minute pores or stomata on the leaves.

In xerophytic plants from hot, dry and almost waterless regions where evaporation would be excessive, the leaf surface, and consequently the number of stomata, are reduced to a minimum, as it would be fatal to such plants to exhale vapour as freely in those regions as the broad-leaved plants that grow in places where there is abundance of moisture. Although transpiration is a necessary accompaniment of nutrition, it may easily become excessive, especially where the plant cannot readily recoup itself. In these circumstances "syringing" and "damping down" are of value in cooling the temperature of the air in hothouses and greenhouses and increasing its humidity, thereby checking excessive transpiration. Shading the glass with canvas or washes during the summer months has the same object in view. Syringing is also beneficial in washing away dirt and insects.

BUDS

The recognition of the various forms of buds and their modes of disposition in different plants is a matter of the first consequence in the operations of pruning and training. Flowerbuds are produced either on the old wood, i.e. the shoots of the past year's growth, or on a shoot of the present year. The peach, horse-chestnut, lilac, morello cherry, black currant, rhododendron and many other trees and shrubs develop flower-buds for the next season speedily after blossoming, and these may be stimulated into premature growth. The peculiar short, stunted branches or "spurs" which bear the flower-buds of the pear, apple, plum, sweet cherry, red currant, laburnum, deserve special attention. In the rose, passion-flower, clematis, honeysuckle, in which the flowerbuds are developed at the ends of the young shoot of the year, we have examples of plants destitute of flower-buds during the winter.

Propagation by Buds

The detached leaf-buds (*gemmae or bulbils*, of some plants are capable under favourable conditions of forming new plants. The edges of the leaves of *Bryophyllum calycinum* and of *Cardamine pratensis*, and the growths in the axils of the leaves of Lilium bulbiferum, as well as the fronds of certain ferns (e.g. *Asplenium bulbiferum*), produce buds of this character. It is a matter of familiar observation that the ends of the shoots of brambles take root when bent down to the ground. In some instances buds form on the roots, and may be used for purposes of propagation, as in the Japan quince, the globe thistle, the sea holly, some sea lavenders, *Bocconia, Acanthus*.

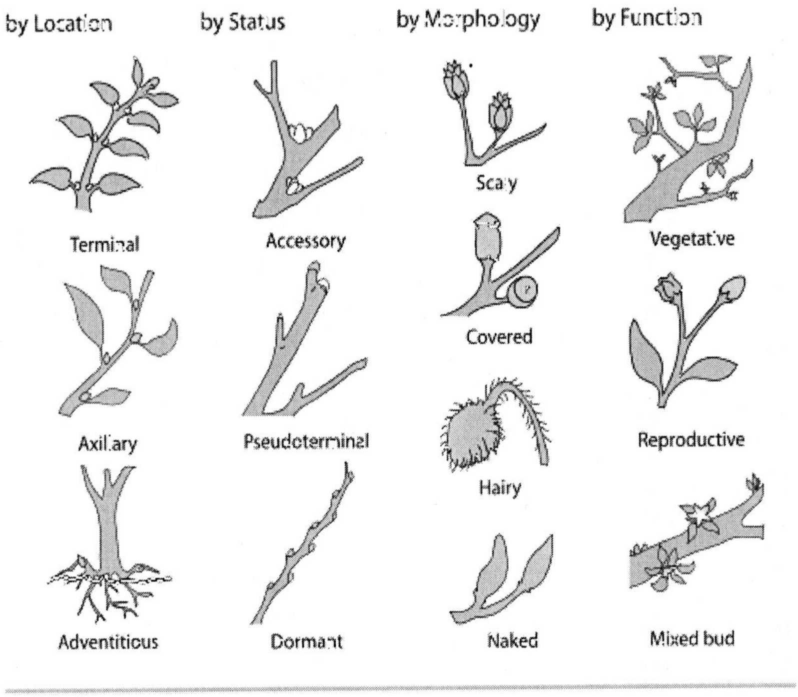

Figure 2. Types of Plant Buds

Of the tendency in buds to assume an independent existence gardeners avail themselves in the operations of striking "cuttings," and making "layers" and "pipings," as also in budding and grafting. In taking a slip or cutting the gardener removes from the parent plant a shoot having one or more buds or "eyes," in the case of the vine one only, and places it in a moist and sufficiently warm situation, where, as previously mentioned, undue evaporation from the surface is prevented.

For some cuttings, pots filled with light soil, with the protection of the propagating-house and of bell-glasses, are requisite; but for many of our hardy deciduous trees and shrubs no such precautions are necessary, and the insertion of a short shoot about half its length into moist and gritty ground at the proper season suffices to ensure its growth. In the case of the more delicate plants, the formation of roots is preceded by the production from the cambium of the cuttings of a succulent mass of tissue, the callus.

It is important in some cases, e.g. zonal pelargoniums, fuchsias, shrubby calceolarias, dahlias, carnations, to retain on the cutting some of its leaves, so as to supply the requisite food for storage in the callus. In other cases, where the buds themselves contain a sufficiency of nutritive matter for the young growths, the retention of leaves is not necessary.

The most successful mode of forming roots is to place the cuttings in a mild bottom-heat, which expedites their growth, even in the case of many hardy plants whose cuttings strike roots in the open soil. With some hard-wooded trees, as the common white-thorn, roots cannot be obtained without bottom-heat. It is a general rule throughout plant culture that the activity of the roots shall be in advance of that of the leaves.

Cuttings of deciduous trees and shrubs succeed best if planted early in autumn while the soil still retains the solar heat absorbed during summer. For evergreens August or September, and for greenhouse and stove-plants the spring and summer months, are the times most suitable for propagation by cuttings.

Layering consists simply in bending down a branch and keeping it in contact with or buried to a small depth in the soil until roots are formed; the connexion with the parent plant may then be severed. Many plants can be far more easily propagated thus than by cuttings.

Grafting or "working" consists in the transfer of a branch, the "graft" or "scion," from one plant to another, which latter is termed the "stock." The operation must be so performed that the growing tissues, or cambium-layer of the scion, may fit accurately to the corresponding layer of the stock. In budding, as with roses and peaches, a single bud only is implanted. Inarching is essentially the promotion of the union of a shoot of one plant to that of another of the same or allied species or variety.

The outer bark of each being removed, the two shoots are kept in contact by ligature until union is established, when the scion is completely severed from its original attachments. This operation is varied in detail according to the kind of plant to be propagated, but it is essential in all cases that the affinity between the two plants be near, that the union be neatly effected, and that the ratio as well as the season of growth of stock and scion be similar.

The selection of suitable stocks is a matter still requiring much scientific experiment. The object of grafting is to expedite and increase the formation of flowers and fruit. Strong-growing pears, for instance, are grafted on the quince stock in order to restrict their tendency to form "gross" shoots and a superabundance of wood in place of flowers and fruit. Apples, for the same reason, are "worked" on the "paradise" or "doucin" stocks, which from their influence on the scion are known as dwarfing stocks. Scions from a tree which is weakly, or liable to injury by frosts, are strengthened by engrafting on robust stocks. Lindley has pointed out that, while in Persia, its native country, the peach is probably best grafted on the peach, or on its wild type the almond, in England, where the summer temperature of the soil is much lower than that of Persia, it might be expected, as experience has proved, to be most successful on stocks of the native plum.

The form and especially the quality of fruit is more or less affected by the stock upon which it is grown. The Stanwick nectarine, so apt to crack and not to ripen when worked in the ordinary way, is said to be cured of these propensities by being first budded close to the ground, on a very strong-growing Magnum Bonum plum, worked on a Brussels stock, and by then budding the nectarine on the Magnum Bonum about a foot from the ground. The fruit of the pear is of a higher colour and smaller on the quince stock than on the wild pear; still more so on the medlar. On the mountain ash the pear becomes earlier. The effects produced by stock on scion, and more particularly by scion on stock, are as a rule with difficulty appreciable. Nevertheless, in exceptional cases modified growths, termed "grafthybrids," have been obtained which have been attributed to the commingling of the characteristics of stock and scion.

Of these the most remarkable example is *Cytisus Adami*, a tree which year after year produces some shoots, foliage and flowers like those of the common laburnum, others like those of the very different looking dwarf shrub C. *purpureus*, and others again intermediate between these. We may hence infer that C. *purpureus* was grafted or budded on the common laburnum, and that the intermediate forms are the result of graft-hybridisation. Numerous similar facts have been recorded. Among gardeners the general opinion is against the possibility of graft-hybridisation.

The wonder, however, seems to be that it does not occur more frequently, seeing that fluids must pass from stock to scion, and matter elaborated in the leaves of the scion must certainly to some extent enter the stock. It is clear, nevertheless, from examination that as a rule the wood of the stock and the wood of the scion retain their external characters year by year without change. Still, as in the laburnum just mentioned, in the variegated jasmine and in *Abutilon Darwinii*, in the copper beech and in the horse-chestnut, the influence of a variegated scion has occasionally shown itself in the production from the stock of variegated shoots.

Planting

By removal from one place to another the growth of every plant receives a check. How this check can be obviated or reduced, with regard to the season, the state of atmosphere, and the condition and circumstances of the plant generally, is a matter to be considered by the practical gardener. As to season, it is now admitted with respect to deciduous trees and shrubs that the earlier in autumn planting is performed the better; although some extend it from the period when the leaves fall to the first part of spring, before the sap begins to move. If feasible, the operation should be completed by the end of November, whilst the soil is still warm with the heat absorbed during summer. Attention to this rule is specially important in the case of rare and delicate plants.

Early autumn planting enables wounded parts of roots to be healed over, and to form fibrils, which will be ready in spring, when it is most required, to collect food for the plant. Planting late in spring should, as far as possible, be avoided, for the buds then begin to awaken into active life, and the draught upon the roots becomes great. It has been supposed that because the surface of the young leaves is small transpiration is correspondingly feeble; but it must be remembered, not only that their newlyformed tissue is unable without an abundant supply of sap from the roots to resist the excessive drying action of the atmosphere, but that, in spring, the lowness of the temperature at that season in Great Britain prevents the free circulation of the sap. The comparative dryness of the atmosphere in spring also causes a greater amount of transpiration then than in autumn and winter. Another fact in favour of autumnal planting is the production of roots in winter. The best way of performing transplantation depends greatly on the size of the trees, the soil in which they grow, and the mechanical appliances made use of in lifting and transporting them.

The smaller the tree the more successfully can it be removed. The more argillaceous and the less siliceous the soil the more readily can balls of earth be retained about the roots. All planters lay great stress on the preservation of the fibrils; the point principally disputed is to what extent they can with safety be allowed to be cut off in transplantation. Trees and shrubs in thick plantations, or in sheltered warm places, are ill fitted for planting in bleak and cold situations.

During their removal it is important that the roots be covered, if only to prevent desiccation by the air. Damp days are therefore the best for the operation; the dryest months are the most unfavourable. Though success in transplanting depends much on the humidity of the atmosphere, the most important requisite is warmth in the soil; humidity can be supplied artificially, but heat cannot. Pruning, or the removal of superfluous growths, is practised in order to equalise the development of the different parts of trees, or to promote it in particular directions so as to secure a certain form, and, by checking undue luxuriance, to promote enhanced fertility. In the rose-bush, for instance, in which, as we have seen, the flower-buds are formed on the new wood of the year, pruning causes the old wood to "break," i.e. to put forth a number of new buds, some of which will produce flowers at their extremities.

The manner and the time in which pruning should be accomplished, and its extent, vary with the plant, the objects of the operation, i.e. whether for the production of timber or fruit, the season and various other circumstances. So much judgment and experience does the operation call for that it is a truism to say that bad pruning is worse than none.

The removal of weakly, sickly, overcrowded and gross infertile shoots is usually, however, a matter about which there can be few mistakes when once the habit of growth and the form and arrangement of the buds are known.

Winter pruning is effected when the tree is comparatively at rest, and is therefore less liable to "bleeding" or outpouring of sap. Summer pruning or pinching off the tips of such of the younger shoots as are not required for the extension of the tree, when not carried to too great an extent, is preferable to the coarser more reckless style of pruning. The injury inflicted is less and not so concentrated; the wounds are smaller, and have time to heal before winter sets in.

The effects of badly-executed pruning, or rather hacking, are most noticeable in the case of forest trees, the mutilation of which often results in rotting, canker and other diseases. Judicious and timely thinning so as to allow the trees room to grow, and to give them sufficiency of light and air, will generally obviate the need of the pruning-saw, except to a relatively small extent.

Training is a procedure adopted when it is required to grow plants in a limited area, or in a particular shape, as in the case of many plants of trailing habit. Judicious training also may be of importance as encouraging the formation of flowers and fruit.

Growth in length is mainly in a vertical direction, or at least at the ends of the shoots; and this should be encouraged, in the case of a timber tree, or of a climbing plant which it is desired should cover a wall quickly; but where flowers or fruit are specially desired, then, when the wood required is formed, the lateral shoots may often be trained more or less downward to induce fertility. The refinements of training, as of pruning, may, however, be carried too far; and not unfrequently the symmetrically trained trees of the French excite admiration in every respect save fertility.

Sports or Bud Variations

Here we may conveniently mention certain variations from the normal condition in the size, form or disposition of buds or shoots on a given plant. An inferior variety of pear, for instance, may suddenly produce a shoot bearing fruit of superior quality; a beech tree, without obvious cause, a shoot with finely divided foliage; or a camellia an unwontedly fine flower. When removed from the plant and treated as cuttings or grafts, such sports may be perpetuated. Many garden varieties of flowers and fruits have thus originated. The cause of their production is very obscure.

FORMATION OF FLOWERS

Flowers, whether for their own sake or as the necessary precursors of the fruit and seed, are objects of the greatest concern to the gardener. As a rule they are not formed until the plant has arrived at a certain degree of vigour, or until a sufficient supply of nourishment has been stored in the tissues of the plant. The reproductive process of which thc formation of the flower is the first stage being an exhaustive one, it is necessary that the plant, as gardeners say, should get "established" before it flowers.

Moreover, although the green portions of the flower do indeed perform the same office as the leaves, the more highly coloured and more specialised portions, which are further removed from the typical leaf-form, do not carry on those processes for which the presence of chlorophyll is essential; and the floral organs may, therefore, in a rough sense, be said to be parasitic upon the green parts. A check or arrest of growth in the vegetative organs seems to be a necessary preliminary to the development of the flower.

A diminished supply of water at the root is requisite, so as to check energy of growth, or rather to divert it from leaf-making. Partial starvation will sometimes effect this; hence the grafting of freegrowing fruit trees upon dwarfing stocks, as before alluded to, and also the "ringing" or girdling of fruit trees, i.e. the removal from the branch of a ring of bark, or the application of a tight cincture, in consequence of which the growth of the fruits above the wound or the obstruction is enhanced. On the same principle the use of small pots to confine the roots, root-pruning and lifting the roots, and exposing them to the sun, as is done in the case of the vine in some countries, are resorted to.

A higher temperature, especially with deficiency of moisture, will tend to throw a plant into a flowering condition. This is exemplified by the fact that the temperature of the climate of Great Britain is too low for the flowering, though sufficiently high for the growth of many plants. Thus the Jerusalem artichoke, though able to produce stems and tubers abundantly, only flowers in exceptionally hot seasons.

Forcing

The operation of forcing is based upon the facts just mentioned. By subjecting a plant to a gradually increasing temperature, and supplying water in proportion, its growth may be accelerated; its season of development may be, as it were, anticipated; it is roused from a dormant to an active state. Forcing therefore demands the most careful adjustment of temperature and supplies of moisture and light. Deficiency of light is less injurious than might at first be expected, because the plant to be forced has stored up in its tissues, and available for use, a reserve stock of material formed through the agency of light in former seasons.

The intensity of the colour of flowers and the richness of flavour of fruit are, however, deficient where there is feebleness of light. Recent experiments show that the influence of electric light on chlorophyll is similar to that of sunlight, and that deficiencies of natural light may to some extent be made good by its use. The employment of that light for forcing purposes would seem to be in part a question of expense. The advantage hitherto obtained from its use has consisted in the rapidity with which flowers have been formed and fruits ripened under its influence, circumstances which go towards compensating for the extra cost of production.

Retardation

The art of retarding the period of flowering in certain plants consists, in principle, in the artificial application of cold temperatures whereby the resting condition induced by low winter temperature is prolonged. For commercial purposes, crowns of lily of the valley, tulip and other bulbs, and such deciduous woody plants as lilac and deciduous species of rhododendron, while in a state of rest, are packed in wet moss and introduced into coldstorage chambers, where they may be kept in a state of quiescence, if desired, throughout the following summer.

The temperature of the cold chamber is varied from the freezing-point of water, to a few degrees lower, according to the needs of the plants under treatment. When required for use they are removed to cool sheds to thaw, and are then gradually inured to higher temperatures. The chief advantages of retarded plants are:

(*a*) they may be flowered almost at will;

(*b*) they are readily induced to flower at those times when unretarded plants refuse to respond to forcing.

Coldstorage chambers form a part of the equipment of most of the leading establishments where flowers are grown for market.

Double Flowers

The taste of the day demands that "double flowers" should be largely grown. Though in many instances, as in hyacinths, they are less beautiful than single ones, they always present the advantage of being less evanescent. Under the vague term "double" many very different morphological changes are included. The flower of a double dahlia, e.g. offers a totally different condition of structure from that of a rose or a hyacinth. The double poinsettia, again, owes its so-called double condition merely to the increased number of its scarlet involucral leaves, which are not parts of the flower at all. It is reasonable, therefore, to infer that the causes leading to the production of double flowers are varied.

A good deal of difference of opinion exists as to whether they are the result of arrested growth or of exuberant development, and accordingly whether restricted food or abundant supplies of nourishment are the more necessary for their production. It must suffice here to say that double flowers are most commonly the result of the substitution of brightly-coloured petals for stamens or pistils or both, and that a perfectly double flower where all the stamens and pistils are thus metamorphosed is necessarily barren.

Such a plant must needs be propagated by cuttings. It rarely happens, however, that the change is quite complete throughout the flower, and so a few seeds may be formed, some of which may be expected to reproduce the double-blossomed plants. By continuous selection of seed from the best varieties, and "roguing" or eliminating plants of the ordinary type, a "strain" or race of double flowers is gradually produced.

FORMATION OF SEED

Fertilisation

In fertilisation — the influence in flowering plants of the male-cell in the pollen tube upon the eggcell in the ovule—there are many circumstances of importance horticulturally, to which, therefore, brief reference must be made. Flowers, generally speaking, are either self-fertilised, cross-fertilised or hybridised. Self-fertilisation occurs when the pollen of a given flower affects the egg-cell of the same individual flower.

Cross-fertilisation varies both in manner and degree. In the simplest instances the pollen of one flower fertilises the ovules of another on the same plant, owing to the stamens arriving at maturity in any one flower earlier or later than the pistils. Cross-fertilisation must of necessity occur when the flowers are structurally unisexual, as in the hazel, in which the male and female flowers are monoecious, or separate on the same plant, and in the willow, in which they are dioecious, or on different plants.

A conspicuous example of a dioecious plant is the common aucuba, of which for years only the female plant was known in Britain. When, through the introduction of the male plant from Japan, its fertilisation was rendered possible, ripe berries, before unknown, became common ornaments of the shrub.

The conveyance of pollen from one flower to another in crossfertilisation is effected naturally by the wind, or by the agency of insects and other creatures. Flowers that require the aid of insects usually offer some attraction to their visitors in the shape of bright colour, fragrance or sweet juices. The colour and markings of a flower often serve to guide the insects to the honey, in the obtaining of which they are compelled either to remove or to deposit pollen.

The reciprocal adaptations of insects and flowers demand attentive observation on the part of the gardener concerned with the growing of grapes, cucumbers, melons and strawberries, or with the raising of new and improved varieties of plants. In wind-fertilised plants the flowers are comparatively inconspicuous and devoid of much attraction for insects; and their pollen is smoother and smaller, and better adapted for transport by the wind, than that of insectfertilised plants, the roughness of which adapts it for attachment to the bodies of insects. It is very probable that the same flower at certain times and seasons is self-fertilising, and at others not so.

The defects which cause gardeners to speak of certain vines as "shy setters," and of certain strawberries as "blind," may be due either to unsuitable conditions of external temperature, or to the non-accomplishment, from some cause or other, of cross-fertilisation. In a vinery, tomatohouse or a peach-house it is often good practice at the time of flowering to tap the branches smartly with a stick so as to ensure the dispersal of the pollen. Sometimes more delicate and direct manipulation is required, and the gardener has himself to convey the pollen from one flower to another, for which purpose a small camel's-hair pencil is generally suitable.

The degree of fertility varies greatly according to external conditions, the structural and functional arrangements just alluded to, and other causes which may roughly be called constitutional. Thus, it often happens that an apparently very slight change in climate alters the degree of fertility. In a particular country or at certain seasons one flower will be self-sterile or nearly so, and another just the opposite.

Hybridisation

Some of the most interesting results and many of the gardener's greatest triumphs have been obtained by hybridisation, i.e. the crossing of two individuals not of the same but of two distinct species of plants, as, for instance, two species of rhododendron or two species of orchid. It is obvious that hybridisation differs more in degree than in kind from cross-fertilisation. The occurrence of hybrids in nature explains the difficulty experienced by botanists in deciding on what is a species, and the widely different limitations of the term adopted by different observers in the case of willows, roses, brambles.

The artificial process is practically the same in hybridisation as in cross-fertilisation, but usually requires more care. To prevent self-fertilisation, or the access of insects, it is advisable to remove the stamens and even the corolla from the flower to be impregnated, as its own pollen or that of a flower of the same species is often found to be "prepotent."

There are, however, cases, e.g. some passion-flowers and rhododendrons, in which a flower is more or less sterile with its own, but fertile with foreign pollen, even when this is from a distinct species. It is a singular circumstance that reciprocal crosses are not always or even often possible; thus, one rhododendron may afford pollen perfectly potent on the stigma of another kind, by the pollen of which latter its own stigma is unaffected.

The object of the hybridiser is to obtain varieties exhibiting improvements in hardihood, vigour, size, shape, colour, fruitfulness, resistance to disease or other attributes. His success depends not alone on skill and judgment, for some seasons, or days even, are found more propitious than others. Although promiscuous and hap-hazard procedures no doubt meet with a measure of success, the best results are those which are attained by systematic work with a definite aim. Hybrids are sometimes less fertile than pure-bred species, and are occasionally quite sterile. Some hybrids, however, are as fertile as pure-bred plants. Hybrid plants may be again crossed, or even re-hybridised, so as to produce a progeny of very mixed parentage. This is the case with many of our roses, dahlias, begonias, pelargoniums, orchids and other long or widely cultivated garden plants.

Reversion

In modified forms of plants there is frequently a tendency to "sport" or revert to parental or ancestral characteristics. So markedly is this the case with hybrids that in a few generations all traces of a hybrid origin may disappear. The dissociation of the hybrid element in a plant must be obviated by careful selection. The researches of Gregor Johann Mendel, abbot of the Augustinian monastery at Briinn, in connexion with peas and other plants, apparently indicate that there is a definite natural law at work in the production of hybrids. Having crossed yellow and green seeded peas both ways, he found that the progeny resulted in all yellow coloured seeds. These gave rise in due course to a second generation in which there were three yellows to one green.

In the third generation the yellows from the second generation gave the proportion of one pure yellow, two impure yellows, and one green; while the green seed of the second generation threw only green seeds in the third, fourth and fifth generations. The pure yellow in the third generation also threw pure yellows in the fourth and fifth and succeeding generations. The impure yellows, however, in the next generation gave rise to one pure yellow, one pure green, to two impure yellows, and so on from generation to generation.

Germination

The length of the period during which seeds remain dormant after their formation is very variable. The conditions for germination are much the same as for growth in general. Access to light is not required, because the seed contains a sufficiency of stored-up food. The temperature necessary varies according to the nature and source of the seed. Some seeds require prolonged immersion in water to soften their shells; others are of so delicate a texture that they would dry up and perish if not kept constantly in a moist atmosphere.

Seeds buried too deeply receive a deficient supply of air. As a rule, seeds require to be sown more deeply in proportion to their size and the lightness of the soil. The time required for germination in the most favourable circumstances varies very greatly, even in the same species, and in seeds taken from one pod. Thus the seeds of *Primula japonica*, though sown under precisely similar conditions, yet come up at very irregular intervals of time. Germination is often slower where there is a store of available food in the perisperm, or in the endosperm, or in the embryo itself, than where this is scanty or wanting. In the latter case the seedling has early to shift for itself, and to form roots and leaves for the supply of its needs.

Selection

Supposing seedlings to have been developed, it is found that a large number of them present considerable variations, some being especially robust, others peculiar in size or form. Those most suitable for the purpose of the gardener are carefully selected for propagation, while others not so desirable are destroyed; and thus after a few generations a fixed variety, race or strain superior to the original form is obtained.

Many garden plants have originated solely by selection; and much has been done to improve our breeds of vegetables, flowers and fruit by systematic selection. Large and well-formed seeds are to be preferred for harvesting. The seeds should be kept in sacks or bags in a dry place, and if from plants which are rare, or liable to lose their vitality, they are advantageously packed for transmission to a distance in hermetically sealed bottles or jars filled with earth or moss, without the addition of moisture. It will have been gathered from what has been said that seeds cannot always be depended on to reproduce exactly the characteristics of the plant which yielded them; for instance, seeds of the greengage plum or of the Ribston pippin will produce a plum or an apple, but not these particular varieties, to perpetuate which grafts or buds must be employed.

REFERENCES

Eames, Arthur Johnson and MacDaniels, Laurence H. 1947. *An Introduction to Plant Anatomy*. McGraw-Hill, New York;

Esau, Katherine. 1965. *Plant Anatomy* (2nd ed.) Wiley, New York.

Farabee, M.J. 2001. *Plants and their structure*. Estrella Mountain Community College, Phoenix, Arizona.

Raven, P. H., R. F. Evert, & S. E. Eichhorn. 2005. *Biology of Plants*, 7th ed., W. H. Freeman. New York.

3

Basics of Plant Nutrition

Plants convert light energy into biomass through photosynthesis and produce various products of economic value among others. To do this, plants need sufficient light, suitable temperature, substances such as water, CO_2, oxygen, and a number of nutrients. The survival and well-being of humans and animals depends on plant production, which in turn depends heavily on the availability of mineral and other nutrients. This is why plants and animals have several essential nutrients in common.

Like all organisms, higher green plants need nutrients for their growth and development. Nutrients are indispensable as plant constituents, for biochemical reactions, and for the production of organic materials referred to as photosynthates by photosynthesis. In agriculture, optimal crop nutrition is an important prerequisite for obtaining high yields and good-quality produce. The nutrients required are obtained by plants both from soil reserves and external nutrient sources. Almost all of the 90 natural elements can be found in green plants although most of them have no function.

PLANT NUTRIENTS

Essential Plant Nutrients

A total of only 16 elements are essential for the growth and full development of higher green plants according to the criteria laid down by Arnon and Stout. These criteria are:

- A deficiency of an essential nutrient makes it impossible for the plant to complete the vegetative or reproductive stage of its life cycle.
- Such deficiency is specific to the element in question and can be prevented or corrected only by supplying this element.
- The element is involved directly in the nutrition of the plant quite apart from its possible effects in correcting some unfavourable microbiological or chemical condition of the soil or other culture medium.

Out of these 16 elements, carbon (c) and oxygen are obtained from the gas CO_2, and hydrogen (H) is obtained from water (H_2O). These three elements are required in large quantities for the production of plant constituents such as cellulose or starch. The other 13 elements are called mineral nutrients because they are taken up in mineral (inorganic) forms. They are traditionally divided into two groups, macronutrients and micronutrients, according to the amounts required. Regardless of the amount required, physiologically, all of them are equally important. The 13 mineral elements are taken up by plants in specific chemical forms regardless of their source. Oxygen, C and H make up 95 percent of plant biomass, and the remaining 5 percent is made up by all other elements. The difference in plant concentration between macronutrients and micronutrients is enormous. The relative contents of N and molybdenum (Mo) in plants is in the ratio of 10 000:1. Plants need about 40 times more magnesium (Mg) than Fe. These examples indicate the significant difference between macronutrients and micronutrients.

Beneficial Nutrients

Several elements other than the essential nutrients have beneficial functions in plants. Although not essential, beneficial nutrients can improve the growth of some crops in some respects. Some of these nutrients can be of great practical importance and may require external addition:

- *Nickel (Ni):* a part of enzyme urease for breaking urea in the soil, imparts useful role in disease resistance and seed development.
- *Sodium (Na):* for beets, partly able to replace K (uptake as Na^+).
- *Cobalt (Co):* for N fixation in legumes and for other plants (uptake as Co^{2+}).
- *Silicon (Si):* for stalk stability of cereals particularly rice (uptake as silicate anion).
- *Aluminium (Al):* for tea plants (uptake as Al^{3+} or similar forms).

As humans and domestic animals require several nutrients in addition to those required by plants, these additional nutrients should also be considered in food or feed production, and their deficiencies corrected by appropriate inputs. In addition to plant nutrients, the elements essential for humans and domestic animals are: Cobalt (Co), selenium (Se), chromium (Cr) and iodine (I).

FUNCTIONS, MOBILITY IN PLANTS AND DEFICIENCY/ TOXICITY SYMPTOMS

Some knowledge of the properties and functions of plant nutrients is helpful for their efficient management and, thus, for good plant growth and high yields. Available nutrients in the soil solution can be taken up by the roots, transported to the leaves and used according to their functions in plant metabolism. Nutrient ions are of extremely small size, i.e. like atoms. There are more than 100 000 million K^+ cations within a single leaf cell and more than 1 000 000 molybdate anions, the micronutrient required in the smallest amount.

In general, N and K make up about 80 percent of the total mineral nutrients in plants; P, S, Ca and Mg together constitute 19 percent, while all the micronutrients together constitute less than 1 percent Most plant nutrients are taken up as positively or negatively charged ions from the soil solution. However, some nutrients may be taken up as entire molecules, e.g. boric acid and amino acids, or organic complexes such as metal chelates and to a very small extent urea. Whether the original sources of nutrient ions in the soil solution are from organic substances or inorganic fertilisers, ultimately, the plants absorb them only in mineral forms.

Plants exhibit many shades of greenness but a medium to dark green colour is usually considered a sign of good health and active growth. Chlorosis or yellowing of leaf colour can be a sign of a marginal deficiency and is often associated with retarded growth. Chlorosis is a light green or rather yellowish discoloration of the whole or parts of the leaf caused by a lower content of chlorophyll. Because the cells remain largely intact, the chlorotic symptoms are reversible, i.e. leaves can become green again after the missing nutrient is added.

A severe deficiency results in death of the tissue. Necrosis is a brownish discoloration caused by decaying tissue, which is destroyed irreversibly. Necrotic leaves cannot be recovered by addition of the missing nutrient, but the plant may survive by forming new leaves. Deficiency symptoms can serve as a guide for diagnosing limiting nutrients and the need for corrective measures. However, chlorotic and necrotic leaves might also result from the toxic effects of nutrients, pollution and also from disease and insect attacks. Therefore, confirmation of the cause is important before corrective measures are taken.

Nitrogen

N is the most abundant mineral nutrient in plants. It constitutes 2–4 percent of plant dry matter. Apart from the process of N fixation that occurs in legumes, plants absorb N either as the nitrate ion (NO_3^-) or the ammonium ion (NH_4^+). N is a part of the chlorophyll and is an essential constituent of all proteins. It is responsible for the dark green colour of stem and leaves, vigorous growth, branching/tillering, leaf production, size enlargement, and yield formation. Absorbed N is transported through the xylem to the leaf canopy as nitrate ions, or it may be reduced in the root region and transported in an organic form, such as amino acids or amides.

N is mobile in the phloem; as such, it can be re-translocated from older to younger leaves under N deficiency and translocated from leaves to the developing seed or fruit. The principal organic forms of N in phloem sap are amides, amino acids and ureides. Nitrate and ammonium ions are not present in this sap. N deficiency in plants results in a marked reduction in growth rate. N-deficient plants have a short and spindly appearance. Tillering is poor, and leaf area is small.

As N is a constituent of chlorophyll, its deficiency appears as a yellowing or chlorosis of the leaves. This yellowness usually appears first on the lower leaves while upper leaves remain green as they receive some N from older leaves. In a case of severe deficiency, leaves turn brown and die. As a result, crop yield and protein content are reduced. The effects of N toxicity are less evident than those of its deficiency. They include prolonged growing period and delayed crop maturity. High NH_4^+ in solution can be toxic to plant growth, particularly where the solution is alkaline.

The toxicity results from ammonia (NH_3), which is able to diffuse through plant membranes and interfere with plant metabolism. The potential hydrogen (pH – negative log of H^+ concentration) determines the balance between NH_3 and NH_4^+.

Phosphorus

P is much less abundant in plants having a concentration of about one-fifth to one-tenth that of N in plant dry matter. P is absorbed as the orthophosphate ion (either as $H_2PO_4^+$ or HPO_4^{2-}) depending on soil pH. As the soil pH increases, the relative proportion of $H_2PO_4^-$ decreases and that of HPO_4^{2-} increases. P is essential for growth, cell division, root lengthening, seed and fruit development, and early ripening. It is a part of several compounds including oils and amino acids.

The P compounds adenosine diphosphate (ADP) and adenosine triphosphate (ATP) act as energy carriers within the plants. P is readily mobile within the plant both in the xylem and phloem tissues. When the plant faces P shortage, P from the old leaves is readily translocated to young tissue. With such a mobile element, the pattern of redistribution seems to be determined by the properties of the source and the sink. Plant growth is markedly restricted under P deficiency, which retards growth, tillering and root development and delays ripening.

The deficiency symptoms usually start on older leaves. A bluish-green to reddish colour develops, which can lead to bronze tints and red colour. A shortage of inorganic phosphate in the chloroplast reduces photosynthesis. Because ribonucleic acid (RNA) synthesis is reduced, protein synthesis is also reduced. A decreased shoot/root ratio is a feature of P deficiency, as is the overall lower growth of tops. Extremely high levels of P can result in toxicity symptoms. These generally manifest as a watery edge on the leaf tissue, which subsequently becomes necrotic. In very severe cases, P toxicity can result in the death of the plant.

Potassium

K is the second most abundant mineral nutrient in plants after N. It is 4–6 times more abundant than the macronutrients P, Ca, Mg and S. K is absorbed as the monovalent cation K^+ and it is mobile in the phloem tissue of the plants. K is involved in the working of more than 60 enzymes, in photosynthesis and the movement of its products to storage organs, water economy and providing resistance against a number of pests, diseases and stresses. It plays a role in regulating stomatal opening and, therefore, in the internal water relations of plants. The general symptom of K deficiency is chlorosis along the leaf boundary followed by scorching and browning of tips of older leaves. The affected area moves inwards as the severity of deficiency increases.

K-deficiency symptoms show on the older tissues because of the mobility of K. Affected plants are generally stunted and have shortened internodes. Such plants have: slow and stunted growth; weak stalks and susceptibility to lodging; greater incidence of pests and diseases; low yield; shrivelled grains; and, in general, poor crop quality. Slow plant growth can be accompanied by a higher rate of respiration, which means a wasteful consumption of water per unit of dry matter produced. Kdeficient plants may lose control over the rate of transpiration and suffer from internal drought.

Calcium

Calcium (Ca) ranks with Mg, P and S in the group of least abundant macronutrients in plants. It is absorbed by plant roots as the divalent cation Ca^{2+}. Ca is a part of the architecture of cell walls and membranes. It is involved in cell division, growth, root lengthening and activation or inhibition of enzymes. Ca is immobile in the phloem. Ca deficiency is seen first on growing tips and the youngest leaves. This is the case with all nutrients that are not very mobile in the plants. The Ca-deficiency problems are often related to the inability of Ca to be transported in the phloem. The problems occur in organs that do not transpire readily, i.e. large, fleshy developing fruits. Ca-deficient leaves become small, distorted, cup-shaped, crinkled and dark green. They cease growing, become disorganised, twisted and, under severe deficiency, die. Although all growing points are sensitive to Ca deficiency, those of the roots are affected more severely. Groundnut shells may be hollow or poorly filled as a result of incomplete kernel development.

Magnesium

Mg ranks with Ca, P and S in the group of least abundant macronutrients in plants. Plants take up Mg in the form of Mg^{2+}. Mg occupies the centre-spot in the chlorophyll molecule and, thus, is vital for photosynthesis. It is associated with the activation of enzymes, energy transfer, maintenance of electrical balance, production of proteins, metabolism of carbohydrates, etc. Mg is mobile within the plants. As Mg is readily translocated from older to younger plant parts, its deficiency symptoms first appear in the older parts of the plant.

A typical symptom of Mg deficiency is the interveinal chlorosis of older leaves in which the veins remain green but the area between them turns yellow. As the deficiency becomes more severe, the leaf tissue becomes uniformly pale, then brown and necrotic. Leaves are small and break easily. Twigs become weak and leaves drop early. However, the variety of symptoms in different plant species is so great that their generalised description is more difficult in case of Mg than for other nutrients.

Sulphur

S is required by crops in amounts comparable with P. The normal total S concentration in vegetative tissue is 0.12–0.35 percent and the total N/total S ratio is about 15. Plant roots absorb S primarily as the sulphate ion (SO_4^{2-}). However, it is possible for plants to absorb sulphur dioxide (SO_2) gas from the atmosphere at low concentrations. S is a part of amino acids cysteine, cystine and methionine. Hence, it is essential for protein production. S is involved in the formation of chlorophyll and in the activation of enzymes. It is a part of the vitamins biotin and thiamine (B_1), and it is needed for the formation of mustard oils, and the sulphydryl linkages that are the source of pungency in onion, oils, etc. S moves upwards in the plant as inorganic sulphate anion (SO_4^{2-}). Under low S conditions. mobility is low as the S in structural compounds cannot be translocated.

As the S status of the plant rises, so does its mobility. This pattern of mobility means that in plants with adequate S, sulphate is preferentially translocated to young, actively growing leaves. As the supply of S becomes more limiting, young leaves lack S and, hence, show deficiency symptoms. In many ways, S deficiency resembles that of N. It starts with the appearance of pale yellow or light-green leaves. Unlike N deficiency, S-deficiency symptoms in most cases appear first on the younger leaves, and are present even after N application.

Plants deficient in S are small and spindly with short and slender stalks. Their growth is retarded, and maturity in cereals is delayed. Nodulation in legumes is poor and N fixation is reduced. Fruits often do not mature fully and remain light green in colour. Oilseed crops deficient in S produce a low yield and the seeds have less oil in them. S toxicity can occur under highly reduced conditions, possibly as a result of sulphide (H_2S) injury. Most plants are susceptible to high levels of atmospheric SO_2. Normal SO_2

concentrations range from 0.1 to 0.2 mg SO_2/m^3, and toxicity symptoms are observed when these exceed 0.6 mg SO_2/m^3. S-toxicity symptoms appear as necrotic spots on leaves, which then spread over the whole leaf.

Boron

Boron (B) is probably taken up by plants as the undissociated boric acid (H_3BO_3). It appears that much of the B uptake mainly follows water flow through roots. B in a plant is like the mortar in a brick wall, the bricks being the cells of growing parts such as tips (meristems). Key roles of B relate to: (i) membrane integrity and cell-wall development, which affect permeability, cell division and extension; and (ii) pollen tube growth, which affects seed/fruit set and, hence, yield. B is relatively immobile in plants and, frequently, the B content increases from the lower to the upper parts of plants.

B deficiency usually appears on the growing points of roots, shoots and youngest leaves. Young leaves are deformed and arranged in the form of a rosette. There may be cracking and cork formation in the stalks, stem and fruits; thickening of stem and leaves; shortened internodes, withering or dying of growing points and reduced bud, flower and seed production.

Other symptoms are: premature seed drop or fruit drop; crown and heart rot in sugar beet; hen- and chicken-type bunches in grapes; barren cobs in maize; hollow heart in groundnut; unsatisfactory pollination; and poor translocation of assimilates. Death of the growing tip leads to sprouting of auxiliary meristem and a bushy broom-type growth. Roots become thick, slimy and have brownish necrotic spots. B toxicity can arise under excessive B application, in arid or semi-arid areas, and where irrigation water is rich in B content. B-toxicity symptoms are yellowing of the leaf tip followed by gradual necrosis of the tip and leaf margins, which spreads towards the midrib. Leaves become scorched and may drop early.

Chlorine

Chlorine (Cl) is absorbed as the chloride anion (Cl^-). It is thought to be involved in the production of oxygen during photosynthesis, in raising cell osmotic pressure and in maintaining tissue hydration. Some workers consider it essential only for palm and kiwi fruit. Deficiency of Cl leads to chlorosis in younger leaves and overall wilting as a consequence of the possible effect on transpiration. Cl^-toxicity symptoms are: burning of the leaf tips or margins; bronzing; premature yellowing; leaf fall; and poor burning quality of tobacco.

Copper

Copper (Cu) is taken up as Cu^{2+}. Its uptake appears to be a metabolically mediated process. However, Cu uptake is largely independent of competitive effects and relates primarily to the levels of available Cu in the soil. Cu is involved in chlorophyll formation and is a part of several enzymes such as cytochrome oxidase. As much as 70 percent of the Cu in plants may be present in the chlorophyll, largely bound to chloroplasts. It participates in lignin formation, protein and carbohydrate metabolism, and is possibly required for symbiotic N fixation.

Cu is a part of plastocyanin, which forms a link in the electron transport chain involved in photosynthesis. Cu is not readily mobile in the plant and its movement is strongly dependent on the Cu status of the plant. Cu-deficiency symptoms are first visible in the form of narrow, twisted leaves and pale white shoot tips. At maturity, panicles/ ears are poorly filled and even empty where the deficiency is severe. In fruit trees, dieback of the terminal growth can occur. In maize, yellowing between leaf veins takes place, while in citrus the leaves appear mottled and there is dieback of new twigs. Cu-toxicity symptoms are more variable with species and less established than its deficiency symptoms. Excess Cu induces Fe deficiency and, therefore, chlorosis is a common symptom.

Iron

Fe is absorbed by plant roots as Fe^{2+}, and to a lesser extent as Fe chelates. For efficient utilisation of chelated Fe, separation between Fe and the organic ligand has to take place at the root surface, after the reduction of Fe^{3+} to Fe^{2+}. Absorbed Fe is immobile in the phloem. Fe is generally the most abundant of the micronutrients with a dry-matter concentration of about 100 µg/g (ppm). It plays a role in the synthesis of chlorophyll, carbohydrate production, cell respiration, chemical reduction of nitrate and sulphate, and in N assimilation. Fe deficiency begins to appear on younger leaves first.

Otherwise, its deficiency symptoms are somewhat similar to those of Mn, as both Fe and Mn lead to failure in chlorophyll production. Yellowing of the interveinal areas of leaves (commonly referred to as iron chlorosis) occurs. In severe deficiency, leaves become almost pale white because of the loss of chlorophyll. In cereals, alternate yellow and green stripes along the length of the leaf blade may be observed. Complete leaf fall can occur and shoots can die. Fe toxicity of rice is known as bronzing. In this disorder, the leaves are first covered by tiny brown spots that develop into a uniform brown colour. It can be a problem in highly reduced rice soils as flooding may increase the levels of soluble Fe from 0.1 to 50–100 µg/g Fe within a few weeks. It can also be a problem in highly weathered, lowland acid soils.

Manganese

Manganese (Mn) is taken up by plants as the divalent ion Mn^{2+}. It is known to activate several enzymes and functions as an auto-catalyst. It is essential for splitting the water molecule during photosynthesis. It has certain properties similar to Mg. It is also important in N metabolism and in CO_2 assimilation. Like Fe, it is generally immobile in the phloem. Mn-deficiency symptoms resemble those of Fe and Mg deficiency where interveinal chlorosis occurs in the leaves. However, Mn-deficiency symptoms are first visible on the younger leaves whereas in Mg deficiency, the older leaves are affected first.

Mn deficiency in oats is characterised by "grey-speck" where the leaf blade develops grey lesions but the tip remains green, the base dies and the panicle may be empty. In dicots, younger leaves develop chlorotic patches between the veins. Mn-toxicity symptoms lead to the development of brown spots, mainly on older leaves and uneven green colour. Some disorders caused by Mn toxicity are: crinkle leaf spot in cotton; stem streak; necrosis of potato; and internal bark necrosis of apple trees.

Molybdenum

Mo is absorbed as the molybdate anion MoO_4^{2-} and its uptake is controlled metabolically. Mo is involved in several enzyme systems, particularly nitrate reductase, which is needed for the reduction of nitrate, and nitrogenase, which is involved in BNF. Thus, it is involved directly in protein synthesis and N fixation by legumes. Mo appears to be moderately mobile in the plant. This is suggested by the relatively high levels of Mo in seeds, and because deficiency symptoms appear in the middle and older leaves. Mo deficiency in legumes can resemble N deficiency because of its role in N fixation. Mo deficiency can cause marginal scorching and rolling or cupping of leaves and yellowing and stunting in plants. Yellow spot disease in citrus and whip tail in cauliflower are commonly associated with Mo deficiency. Fodders containing more than 5 µg/g Mo in the dry matter are suspected to contain toxic levels of Mo for grazing animals.

Zinc

Zn is taken up as the divalent cation Zn^{2+}. Early work suggested that Zn uptake was passive, but more recent work indicates that it is active. Zn is required directly or indirectly by several enzymes systems, auxins and in protein synthesis, seed production and rate of maturity. Zn is believed to promote RNA synthesis, which in turn is needed for protein production. The mobility of Zn is low.

The rate of Zn mobility to younger tissue is particularly depressed in Zn-deficient plants. Common symptoms of Zn deficiency are: stunted plant growth; poor tillering; development of light green, yellowish, bleached spots; chlorotic bands on either side of the midrib in monocots; brown rusty spots on leaves in some crops, which in acute Zn deficiency as in rice may cover the lower leaves; and in fruit trees the shoots may fail to extend and the small leaves may bunch together at the tip in a rosette-type cluster. Little-leaf condition is also a common symptom. Internodes are short. Flowering, fruiting and maturity can be delayed.

Shoots may die off and leaves can fall prematurely. Deficiency symptoms are not the same in all plants. Zn toxicity can result in reduction in root growth and leaf expansion followed by chlorosis. It is generally associated with tissue concentrations greater than 200 μg/g Zn.

BENEFICIAL ELEMENTS

Nickel

Ni is a part of the enzyme urease, which breaks down urea in the soil. It also plays a role in imparting disease resistance and is considered essential for seed development. Information on various aspects of Ni as a micronutrient is gradually becoming available.

Silicon

Si is taken up as the undissociated $Si(OH)_4$ monosilicic acid. The prevalent form of Si in plants is silica gel in the form of hydrated amorphous silica (SiO_2 in H_2O), or polymerised silicic acid, which is immobile in the plant. The beneficial effects of Si on plants include increases in yield that can result from increasing leaf erectness, decreasing susceptibility to lodging, decreasing incidence to fungal infections, and prevention of Mn and/or Fe toxicity. Thus, Si is able to counteract the effects of high N, which tend to increase lodging. In lowland or wetland rice that is low in Si, vegetative growth and grain production is reduced severely and deficiency symptoms such as necrosis of the mature leaves and wilting can occur. Similarly, sugar cane suffers growth reduction under conditions of low Si availability.

Cobalt

Co is taken up as the divalent cation Co^{2+}. It is essential for N-fixing microorganisms, irrespective of whether they are free-living or symbiotic. Co is the metal component of vitamin B_{12}. Thus, Co deficiency inhibits the formation of leghaemoglobin and, hence, N_2 fixation. The Co content of the shoots can be used as an indicator of Co deficiency in legumes, where the critical levels are between 20 and 40 ppb of shoot dry weight.

BASICS OF PLANT NUTRITION

Plant nutrition is governed by some basic facts and principles concerning nutrient supply, their absorption, transport and production efficiency.

Nutrient Demand and Supply

Plants require nutrients in balanced amounts depending on their stage of development and yield levels. For optimal nutrition of crops, a sufficient concentration of the individual nutrients should be present in the plant leaves at any time. An optimal nutrient supply requires:

- sufficient available nutrients in the rootzone of the soil;
- rapid transport of nutrients in the soil solution towards the root surface;
- satisfactory root growth to access available nutrients;
- unimpeded nutrient uptake, especially with sufficient oxygen present;
- satisfactory mobility and activity of nutrients within the plant.

The nutrient concentrations required in plants, or rather in the active tissues, are usually indicated on a dry-matter basis, as this is more reliable than on a fresh-matter basis with its varying water content. Leaves usually have higher nutrient concentrations than do roots. These are usually stated as a percentage for macronutrients and in micrograms per gram for micronutrients.

The Law of the Minimum and its Implications

In plant nutrition, there is a law known as Liebig's law of the minimum. It is named after its author, Justus von Liebig, who said that the growth of a plant is limited by the nutrient that is in shortest supply. Once its supply is improved, the next limiting nutrient controls plant growth. This concept has been depicted in many ways. One is to imagine a barrel with staves of different heights. Such a barrel can only hold water up to the height of its shortest stave. The barrel can be full only when all its staves are of the same size. A plant can also produce to its full potential when all nutrients are at an optimal level, i.e. without any deficiencies or excesses. In order to produce high yields, plant nutrition requires a continuous effort to eliminate minimum factors and provide balanced nutrition in the optimal range. Even if the law of the minimum is only a guiding rule, it serves as a useful basis for nutrient management.

In a broader sense, the law of the minimum can be extended to include all production inputs, not only nutrients. Important aspects of the influence of nutrient supply on plant growth are:

- Plants need certain concentrations of nutrient in their tissue for active growth.
- Nutrient requirement comes somewhat in advance of plant growth.
- Insufficient nutrient uptake results in slight to severe deficiencies.

- Slight deficiencies are not visible and denote "hidden hunger".
- Deficiency symptoms indicate a severe shortage of the nutrient in question.
- High yields are only obtained where all nutrients are in the optimal supply range.
- The nutrient with the lowest supply determines the yield level.
- Many mistakes in fertilisation can be attributed to disregarding the law of the minimum.
- It is easier to correct nutrient deficiencies than to eliminate nutrient toxicities.

Nutrient uptake in time and contents

The pattern of nutrient uptake follows a sigmoid curve in most cases, being first low in the early stages of crop growth, increasing rapidly when drymatter production is maximal and then declining towards crop maturity. During vegetative growth, the daily nutrient uptake increases as growth progresses and reaches a maximum during the main growing period. N, P and K are mainly taken up during active vegetative growth for high photosynthetic activity. The rate of N uptake generally exceeds the rate of drymatter production in the early stages. Phosphate has an additional small peak requirement for early root growth. Modern high-yielding grain varieties continue to absorb P close to maturity and, like N, 70–80 percent of absorbed P ends up in the panicles or ear heads. For fast-growing crops and high yields, the daily nutrient supply must be adequate, especially during the period of maximum requirement. Field crops generally absorb K faster then they absorb N and P. In rice, 75 percent of the K requirement of the plant may be absorbed up to boot leaf stage. Between tillering and panicle initiation, mean daily absorption rates can approach 2.5 kg K_2O/ha/day. Unlike N and P, only 20–25 percent of absorbed K is transferred to the grain, the rest remaining in the straw.

During the final stages of growth as the plant approaches its reproductive phase before maturity, nutrient uptake decreases. Perennial plants retrieve most of the nutrients from the leaves before leaf fall and relocate these for future use. In certain plants, such as jute, a considerable proportion of the absorbed nutrients is returned to the soil through leaf shedding before the crop matures. The highest concentrations of nutrients are found in leaves at early growth stages, and the lowest in leaves near harvest. This decrease in nutrient concentration over time is because of the transfer to other organs and also what is called the dilution effect, which results from a larger increase in dry matter than in nutrient content. For example, young plants with 50 kg K in 1 500 kg of dry matter contain 3.3 percent K but plants approaching flowering with 100 kg K in 5 000 kg of dry matter contain 2.0 percent K. The dilution effect makes the interpretation of plant analysis results difficult, but it can be taken into account by relating plant data to a certain stage of growth.

Nutrient mobility and its effect on deficiency symptoms

While nutrients are transported easily from roots to shoots, their redistribution from the original place of deposition is more difficult for the so-called immobile nutrients. In the event of nutrient deficiency, a partial re-activation is required in order to supply newly formed leaves from the reserves of older ones. The relative mobility of a nutrient within the plant is helpful in understanding the reasons for the differential appearance of nutrient deficiency symptoms as discussed above.

The range of nutrient supply from deficiency to toxicity

The nutrient status of a plant can range from acute deficiency to acute toxicity. A broad division of nutrient status into three groups namely deficient, optimal and excess may be useful for general purposes. For a more accurate assessment of the nutritional status of plants, detailed categorisation is required in which six different ranges can be distinguished:

- *Acute deficiency:* It is associated with definite visible symptoms and poor growth. The addition of the deficient nutrient usually results in increased growth and yields. This range should be avoided as its occurrence is a sign of low nutrient supply or poor nutrient management and poor crop performance.
- *Marginal or latent deficiency*: It is a small range with or without visible deficiency symptoms. However, growth and yield are reduced. Addition of the yield-limiting nutrient results in higher yields but this may not be visible. Optimal nutrient supply prevents hidden hunger.
- *Optimal supply*: This is the range to aim for. Here all nutrients are at the most desired level. In this range, healthy green plants, good growth and high yields with good quality can be expected. This range is generally wide for most nutrients. The optimal supply is reached above the critical concentration, which is generally associated with 90 percent of maximum yield. The critical concentration serves as a diagnostic index for nutrient supply through plant analysis.
- *Luxury supply:* Although there is no definite borderline between optimal and luxury supply, it is useful to identify this range of unnecessarily high nutrient supply. Even if there may not be any negative effects on plant growth or yield, nutrient input is wasted and product quality as well as disease resistance may be reduced especially in the case of excess N. Therefore, luxury consumption of a nutrient should be avoided. In other words, optimal supply should be maintained and not exceeded except in special cases, such as the need for protein enrichment in grain for quality considerations.

- *Marginal or light toxicity:* Here the nutrient concentration is moving towards toxicity. Above the critical toxic concentration, crop growth and yield start to decrease because of the harmful effects of a nutrient surplus, or of toxic substances on biochemical processes and imbalances. No symptoms may be evident, as in the case of hidden hunger.
- *Acute toxicity:* This is the other extreme of excessive supply or poor nutrient management. Plants are damaged by toxic levels resulting in toxicity symptoms, poor or no growth, poor yield, low quality and damage to soil and plant health. The disease resistance of plants may also be lowered and the plant may even die. This range should definitely be avoided for any nutrient.

Nutrient Interactions

It is not easy to provide plants with exactly adequate amounts of all nutrients, and the task is made more difficult by numerous interactions between nutrients. On the one hand, nutrients have their individual specific functions as described above. On the other hand, there are also some common functions as well as interactions. These can be positive or negative. Where a nutrient interaction is synergistic, their combined impact on plant production is greater than the sum of their individual effects where used singly. In an antagonistic interaction, their combined impact on plant production or concentration in tissues is lower than the sum of their individual effects:

- synergistic interaction:
  - effect of nutrient A on yield = 100,
  - effect of nutrient B on yield = 50,
  - effect of combined use of A and B on yield = greater than 150;
- antagonistic interaction:
  - effect of nutrient A on yield = 100,
  - effect of nutrient B on yield = 50,
  - effect of combined use of A and B on yield = lower than 150;
- additive effect:
  - effect of nutrient A on yield = 100,
  - effect of nutrient B on yield = 50,
  - effect of combined use of A and B on yield = 150.

Where they occur, antagonistic interactions are caused mainly by imbalanced nutrient supply and suboptimal nutrient ratios required for satisfactory growth and development. Therefore, from a practical point of view, many unwanted antagonistic interactions can be avoided by maintaining a balanced nutrient supply.

The soundness of a nutrient management programme can be judged from the extent to which it is able to harness the benefits that accrue from positive interactions between nutrients and other production inputs. The synergistic advantage would have been lost and nutrient-use efficiency (NUE) would have been reduced if only one of the two nutrients had been used and the other had been neglected.

Positive interactions have a very high pay-off for farmers, and research must make available all the possible positive interactions for the use of farmers and also tell them how the negative ones can be kept at a safe distance from their fields. The need to harness positive interactions will be felt increasingly as agriculture becomes more intensive and investments in inputs increase.

ROOT GROWTH AND NUTRIENT UPTAKE

As plants absorb nutrient primarily through their roots, regardless of the type of plant, good growth and proliferation of roots is essential for efficient nutrient uptake. Root growth can be favoured or retarded by soil physical and chemical factors. Even small roots must be able to permeate the rooting volume of the soil in both lateral and vertical directions.

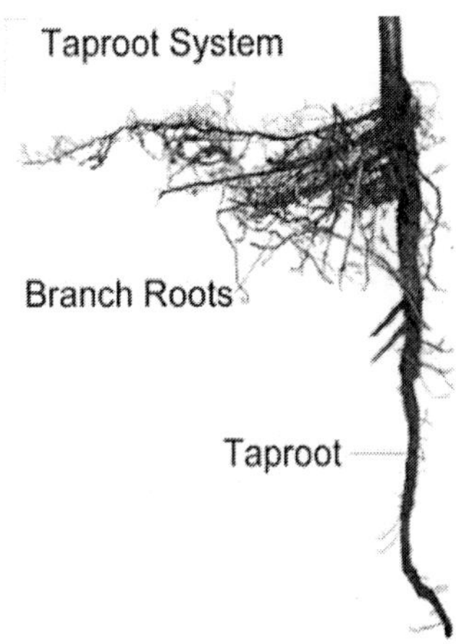

Figure 1. Root system

The major portion of nutrients is taken up by the root hairs, which are about 1–2 mm long and 0.02 mm wide. These are extensions of the epidermal root cells. Root hairs vastly expand the root surface area. Many plants develop several million of these hairs with a total length of more than 10 km. Because very close contact with the soil is required, the amount of fine roots is critical and the number and efficiency of the root hairs is also important.

Uptake of Nutrients from the Soil Solution

The available nutrient forms in the soil are free to move in the soil solution by mass flow or diffusion or up and down the soil profile with water movement. The acquisition of nutrients depends on the size and fineness of the root system, the number of root hairs, the cation exchange capacity (CEC) of the apparent free space (AFS) or the apoplast, etc. A higher CEC results in greater uptake of divalent cations, especially Ca^{2+}. A lower CEC results in greater uptake of monovalent cations such as K^+. The first step in uptake is the entry of the nutrient ion and its passing the outer layer.

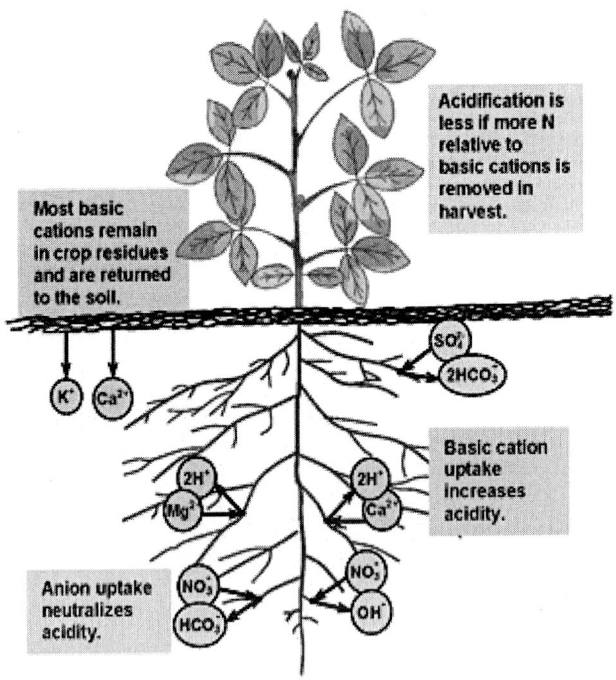

Figure 2. Plant nutrient uptake

Nutrients can enter the cell wall without hindrance. Because of their extremely small size, they are able to penetrate the cell wall tissue of the root hairs. This tissue seems to be a free space and is, therefore, called AFS or the apoplast, a place different from the cytoplast. The second step in nutrient uptake involves movement of the nutrient ion into the cytoplast by crossing the membrane.

The nutrients must be actively taken up into the interior of the cell. The energy required for this uptake is delivered by root respiration, a process that needs oxygen from the soil air and special uptake mechanisms. Thus, nutrient uptake by roots can be active or passive:

- Nutrients can flow passively through the cell wall (AFS) of the root hairs along with the water.
- The free flow ends at the membrane surrounding the active cell substance.
- Nutrients are actively transported through the membrane by special ion carriers.
- Active uptake needs energy from root respiration, which requires sugar and oxygen (O_2).
- Cations are taken up in exchange for H^+ and anions for bicarbonate ions (HCO_3^-) on the root surface.
- Plants can preferentially select nutrients and attempt to exclude unwanted substances.

The fact that nutrient uptake is an active process explains some of its peculiarities. Plants not only accumulate nutrients against a concentration gradient, but they are also able to select from the nutrients at the root surface according to their requirements. In addition, owing to their selection capacity, they can exclude unwanted or even toxic substances, but this exclusion capacity is limited. After uptake into the cytoplast, the nutrients are transported to the next cells and finally arrive at the xylem, which is the tissue through which water and dissolved minerals move upward from the roots to the stem and leaves. They move to the leaves in these water-transporting vessels where they are used for photosynthesis and other processes.

Nutrient uptake by leaves

Apart from gaseous forms of nutrients (CO_2, SO_2, etc.), leaves are able to take up nutrient ions (Fe^{2+}) or even molecules. Although the outer layer of the leaf cuticle closely protects the plant against water loss, nutrients enter the leaves either via the stomata, which serve for gas exchange, or mainly via small micropores of the cuticle and into the apoplast. Foliar application of nutrients is carried out through dilute solutions in order not to damage the leaf cells by osmotic effects.

Most nutrient sources added to the soil involve a monetary expense and, thus, should be utilised, as far as possible, during the vegetative growth period in order to obtain a quick return. Some residual effect during the following season should be acceptable, but losses should be kept low. The magnitude and duration of the residual effect depends on the nutrient, soil properties and cropping intensity. Balanced and adequate supply of plant nutrients is important in order to achieve a high degree of nutrient utilisation by crops, which also results in lower losses. In a wider sense, efficient use of nutrients can only be achieved by considering the whole production system. The nutrition of the plant must be integrated into all aspects of crop management.

REFERENCES

Butt, J. 1993. *Bedding Plant Production Workshop Manual.* Crop Diversification Centre South Publication

Morard, P., A. Pujos, A. Bernadac, and G. Bertoni. 1996. "Effect of temporary calcium deficiency on tomato growth and mineral nutrition". *Journal of Plant Nutrition* 19(1):115-127.

Mohyuddin, M. 1990. *Greenhouse Vegetable Production Guide 1990-91.* Alberta Agriculture Publication. AGDEX 250/15 - 1.

Mirza, M. and M. Younus. 1995. *Plant Nutrition and Fertilizer Management.* Alberta Tree Nursery and Horticulture Centre Publication.

4

Soil Fertility and Crop Production

Soil fertility is a complex quality of soils that is closest to plant nutrient management. It is the component of overall soil productivity that deals with its available nutrient status, and its ability to provide nutrients out of its own reserves and through external applications for crop production. It combines several soil properties, all of which affect directly or indirectly nutrient dynamics and availability. Soil fertility is a manageable soil property and its management is of utmost importance for optimising crop nutrition on both a short-term and a long-term basis to achieve sustainable crop production. Soil productivity is the ability of a soil to support crop production determined by the entire spectrum of its physical, chemical and biological attributes.

Soil fertility is only one aspect of soil productivity but it is a very important one. For example, a soil may be very fertile, but produce only little vegetation because of a lack of water or unfavourable temperature. Even under suitable climate conditions, soils vary in their capacity to create a suitable environment for plant roots. For the farmer, the decisive property of soils is their chemical fertility and physical condition, which determines their potential to produce crops. Good natural or improved soil fertility is essential for successful cropping. It is the foundation on which all input-based high-production systems can be built.

SOILS AS A BASIS FOR CROP PRODUCTION

Soil, the Natural Medium for Plant Growth

Crop production is based largely on soils. For large-scale and low-cost crop production, there is no substitute for natural soils as a substrate for crops in the foreseeable future. Soils are the uppermost part of the earth's crust, formed mainly by the weathering of rocks, formation of humus and by material transfer. Soils vary a great deal in terms of origin, appearance, characteristics and production capacity. Well-developed soils generally show a distinct profile with different layers. The uppermost layer, called topsoil or A horizon, is richest in organic matter, nutrients and various soil organisms. Plants mainly use the topsoil as rooting volume to obtain water and nutrients, but they can also use the subsoil or even lower layers up to 1 m or even deeper. (Fig. 1)

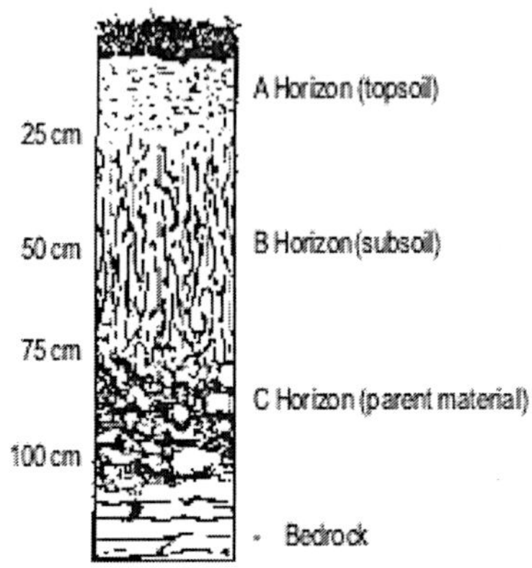

Figure 1: A vertical cross-section of a typical soil profile showing soil horizons (A + B = solum)

Major types of soils are formed from rocks by weathering processes over long periods extending to more than 1000 years. During weathering, physical disintegration of rocks and minerals occurs, and chemical and/or biochemical soil forming processes lead to their decomposition. The result is the synthesis of new products, e.g. clay minerals and humic substances. Mineral or organic substances can be moved downwards or upwards within the profile, but they may also be lost by transportation to other places by water and wind erosion.

Some of the most productive soils are the result of distant long-term geological soil erosion. Soils vary largely with respect to their natural fertility and productivity resulting in plant growth ranging from practically zero to abundant luxuriant growth of natural vegetation. However, only a small proportion of world's soils has a very good level of fertility.

Most soils have only good to medium fertility and some have very low fertility, and are often referred to as marginal soils. Such areas should not generally be used for cropping but only for grazing in a controlled manner. However, under natural vegetation in a suitable climate, even soils of poor fertility may produce luxuriant vegetation where the nutrient cycle is closed, e.g. the Amazon forests. Well-known fertile soils are deep alluvial soils formed from river mud, organicmatter-rich soils on loess material, nutrient rich Vertisols and volcanic soils. In most countries with large food demand, cropping cannot be restricted to the most fertile soils because of the large population and general shortage of usable land. However, soils with medium fertility can be improved considerably as has been demonstrated in many countries. Naturally poor or degraded soils can also be restored to a satisfactory fertility level. Under poor management, soil fertility can be seriously depleted and soils may become useless for agriculture.

Classification of Soils

Soil scientists classify soils by different classification or taxonomic systems. Formerly, the classifications at national level were based on easily recognisable features and relevant soil properties for cropping. Soil-type names were generally well understood by farmers. Even on a higher classification level, the division into zonal soils, intrazonal soils and azonal soils was easy to understand. Modern and global-scale classification systems are based on developmental aspects and resulting special soil properties.

The total land surface of the world is covered by the following major soils:

- soils of humid tropics, e.g. Ferralsols (Oxisols), etc.: 20 percent;
- soils of arid regions, e.g. Calcisols (Calcid), etc.: 18 percent;
- mountainous soils, Leptisols (Umbrept): 15 percent; soils of steppe region, e.g. Chernozems (Udolls): 7 percent;
- Podzols (Spodosols) and similar soils: 7 percent;
- clay soils of subtropics, Vertisols (Vertisols): 3 percent.

As agriculture develops, the natural properties of soils, especially of the topsoils, become more and more similar and adapted to crop requirements. This means that most cropped soils tend to become Arthrosols.

SOIL CONSTITUENTS

A soil consists of mineral matter, organic matter and pore space, which is shared by air, water and life forms. In addition to the above constituents, the soil also contains a large and varied population of micro-organism and macro-organisms that play an important role in plant nutrition. About 45–50 percent of the volume of a normal soil consists of mineral matter, 1–5 percent is organic matter and the remaining 50 percent consists of open pore spaces that are shared by air and water. In a very dry soil, most of these pores are full of air, while in a saturated soil, they are filled with water. Ideally, air and water occupy about equal space, the air residing in the larger pores and water in the smaller ones. Both are needed for the soil to serve as a medium for plant growth. The organic matter and the pores also house a variety of plant and animal life ranging from microscopic bacteria to earthworms and rodents. Of the various soil components, the mineral matter changes little during a farmer's lifetime. The organic matter can be increased, maintained or depleted depending upon the amounts of organic manures used and the rate at which these are decomposed. The áir–water status can change on a day-to-day or even hourly basis.

Soil mineral matter

The soil mineral matter ranges from large pieces of gravel, pebbles and nodules to small grains of sand, silt and clay particles. In addition, there are various oxides, sulphates, silicates and carbonates. The mineral matter on the earth's surface is made up largely of oxygen and Si. Only eight elements are present in amounts greater than 1 percent. Among the plant nutrients, the most abundant are not N or P but Fe, Ca, K and Mg.

The difference in size between a coarse sand particle and fine clay is a thousandfold. The small size of clay particles gives a very large reactive surface area. While all the mineral fractions determine the texture and waterholding capacity (WHC) of a soil, the sand fraction does not do much. The silt acts as a storehouse of certain nutrients, such as K, while the clay fraction is the most active and a class by itself. As described blow, clay particles along with organic matter give the soil its CEC – a property of great importance from a nutrient management point of view. Some common minerals formed in soils are listed below on the basis on their degree of solubility:

- minerals of high solubility:
  - nitrates,
  - chlorides,
  - bicarbonates,
  - sulphates except calcium sulphate;
- minerals of intermediate solubility:
  - gypsum (calcium sulphate),
  - calcite (calcium carbonate),
  - pyrite (under reduced conditions);
- minerals of low solubility:
  - apatite (tricalcium phosphates),
  - oxides and hydroxides of Si, Al and Fe,
  - silicates of Ca, Ma and K.

Saturated solutions of salts with a high solubility inhibit plant growth because of high osmotic pressure. These salts in solid phase are only present in significant quantities under exceptional circumstances. Minerals of intermediate solubility are those with a saturated solution not inhibiting plant growth but their solubility is high enough to contribute significantly to the nutrient composition of the soil solution. The minerals of low solubility contribute to plant nutrient supply only in the long term. Minerals of the silicate, phosphate and oxide groups are almost the end point of the weathering. The fraction of soil mineral matter that contributes to nutrient supply is that which has a moderate degree of water solubility or that with a relatively high specific surface area.

Soil organic matter

Most soils are of mineral origin, but their topsoil contains organic matter that, in spite of its low content, is of great importance to many aspects of soil fertility and plant growth. Soil organic matter (SOM) can range from less than 1 percent in many tropical arid and semi-arid soils of the plains to 5 percent or more in temperate regions or under forest vegetation. The average composition of SOM is 47 percent C, 44 percent O, 7 percent H, 2 percent N and very small amounts of other elements. More than half of SOM consists of carbohydrates, 10–40 percent is the resistant material lignin and the rest consists of compounds of N.

The whole complex of organic matter along with soil organisms and soil flora is of vital importance to soil fertility. SOM contains the well-decomposed fine humus fraction, small plant roots, and members of the plant (flora) and animal kingdoms. SOM plays a role far greater than its share of the soil volume. It is a virtual storehouse of nutrients, plays a direct role in cation exchange and water retention, releases nutrients into the soil solution and produces acids that affect the fixation and release of other nutrients.

SOM or "humus" reaches equilibrium during soil formation. Wet and/or cold soil conditions tend to increase the humus content, whereas high temperatures of tropical climates and cropping procedures promote its decomposition. The C:N ratio provides a general index of the quality of SOM, being in the range of 10–15:1 for fertile soils. When organic manures or green manures are added, these become a part of the organic pool of the soil.

Soil pore space

Soil volume that is not occupied by mineral or organic matter is referred to as pore space. This is shared by soil water, soil air and soil life. It has about ten times more CO_2 than the atmosphere. This CO_2 is produced as a result of breathing by roots and soil micro-organisms. The ratio of pore space to the volume of solid material in the soil is termed the pore space ratio (PSR). It is an important soil property that determines the dynamics of air, water, temperature and nutrients and also the available root space and ease of working the soil.

SOIL PROPERTIES AND PLANT REQUIREMENTS

Plants need anchorage, water and nutrients from the soil but are sensitive to excesses of growth-impeding substances in the soil. The supply and uptake of nutrients from the soil is not a simple process but requires a suitable combination of various soil properties:

- physical properties (depth, texture, structure, pore space with water and air);
- physico-chemical properties;
- chemical and biological properties (nutrient status, their transformation, availability and mobility).

A major objective of having the most suitable soil physical, chemical and biological conditions is to provide the most favourable environment for the roots to grow, proliferate and absorb nutrients.

Soil Physical Properties

Soil physical properties largely determine the texture, structure, physical condition and tilth of the soil. These in turn exert an important influence on potential rooting volume, penetrability of roots, WHC, degree of aeration, living conditions for soil life, and nutrient mobility and uptake. These are as important as soil chemical properties.

Soil depth

Fertile soils generally have a deep rooting zone, which ideally is a minimum of about 1 m for annual crops and 2 m for tree crops. This soil volume should contain no stony or densely compacted layers or unfavourable chemical conditions that impede deep root growth. In addition, the topsoil, which is rich in humus and soil life and the main feeding area for the roots, should be at least 20 cm deep. In practice, many soils have limitations with respect to rooting depth but these can generally be improved by suitable amelioration.

Soil texture

The term texture designates the proportion of different particle size fractions in the soil. Soil texture is primarily represented by its mineral fraction; the organic matter is usually ignored during texture evaluation. Of special importance to soil fertility is the percentage of soil particles of less than 2 mm in diameter, which constitute the fine soil. Fine soil is composed of particles in three size groups: sand, silt and clay. International size units used to classify soil particles in terms of their mean diameter are:

- gravel (> 2.0 mm);
- coarse sand (2.0–0.2 mm);
- fine sand (0.2–0.02 mm);
- silt (0.02–0.002 mm);
- clay (< 0.002 mm).

Based on the relative proportions of these components, soils are classified into different textural classes, such as sandy, loamy or clay soils, and several intermediate classes. The size of a particle is not related to its chemical composition, e.g. sand may be just silica or nutrient rich feldspar or apatite. Clay particles are of colloidal size and are of special importance because of their mechanical and chemical activity. They consist of plate-like structures that have large external and internal charged surfaces for exchange of nutrient ions, particularly the positively charged cations. Some common types of clays are kaolinite, illite and montmorillonite.

The texture of a soil can be determined approximately by simple "finger rubbing" of moist samples, or precisely by conducting a mechanical analysis in a laboratory. Terms used by farmers and sometimes even by researchers such as "light" for sandy soils and "heavy" for clay soils are not based on the actual weight, but on the practical perception of ease of cultivation as mechanical cultivation requires less energy on a light soil than on a heavy soil.

Soil texture influences to a large extent several components of soil fertility such as the amount of nutrient reserves and their proportion to the available nutrient fraction. It also influences several properties such as aeration, pore space distribution, WHC and drainage characteristics. The broad relation of soil texture with soil fertility can be stated as follows:

- Sandy soils are generally poor in nutrient reserves and have a low WHC, but provide favourable conditions for root growth, soil aeration and drainage of surplus water.
- Clay soils are often rich in nutrient reserves (but not necessarily in plant available forms), have high WHC because of the many medium and small pores, but soil aeration is restricted.
- Loamy soils, like sandy or silty loam, have intermediate properties and are generally most suitable for cropping.

A good soil for plant growth should contain moderate quantities of all the different fractions. As the proportion of any fraction increases, such a soil becomes more suitable for plant growth in some respects and less suitable in other respects. Where a soil contains a large proportion of sand, it is well aerated, excessive moisture drains away easily, and the soil is easy to cultivate. However, such a soil retains too little moisture and plants can quickly suffer from moisture stress. Plant nutrients also leach out of such a soil very easily. Where a soil has a high clay content, it retains moisture and plant nutrients well but such a soil is also poorly aerated, becomes easily waterlogged, and is difficult to cultivate. Thus, it is easy to understand why sandy loams are considered the most productive agricultural soils.

Soil structure

The individual particles of the fine soil fraction are usually bound together by organic/inorganic substances into larger aggregates. The process, known as aggregation, results in a vastly increased pore space that is occupied by air and water. The three-dimensional arrangement of the different sizes and shapes of soil aggregates is termed soil structure. In contrast to soil texture, it is a rather variable soil property that, from an agronomic point of view, can improve or deteriorate.

There are different types of soil structure, e.g.: single grain and granular structure with good water permeability; blocky and prismatic structure with medium water permeability; and platy and massive structure with slow water permeability. The clay particles, some of which have swelling and shrinking properties depending on water content are important components of structure formation and, therefore, of pore space distribution.

For agricultural use, the best type is a stable or large granular "crumb" structure with biologically formed sponge-like aggregates of 0.1–1 cm formed by earthworms in combination with string-forming fungal hyphae or gum-producing microbes. Such crumbs are stable against wetting and have a good mixture of different pore sizes, which are desirable characteristics of fertile soils. In contrast, crumbly pieces formed by mechanical tillage are usually much less stable.

An important feature of good soil structure is its stability against deteriorating processes such as wetting and pressure. Annual cropping with relatively high disturbance of soils often results in some structural deterioration, which can be reversed to different degrees.

The soil pore system with water and air

The solid soil particles leave large and small holes between them, which make the soil a porous system. The PSR determines the dynamics of air, water, temperature and nutrients and also the available root space and ease of working the soil. Because of the large portion of pores in soils, the volume weight of mineral topsoils is only about 1.5 (1 litre of soil = 1.5 kg). In a soil, there are a wide range of pore sizes present, and the percentage of the total pore space made up of any particular size varies greatly between soil textural classes. The multiple-shaped pore space is filled by water and air in varying proportions depending on the water content of the soil. Ideally, mineral soils should have a pore volume of almost 50 percent with about one-third of this consisting of large pores. The size of pores determines their function:

- *Large pores:* These contain air or provide drainage.
- *Medium pores (0.2–10 μm diameter)*: These contain the available water.
- *Small pores*: These also hold water, but because of the high tension with which it is tightly held, this water is not available to the plants.

Soil water

Soil water added by rain or irrigation is stored up to the WHC of the soil, which is also called the field capacity. The WHC denotes the maximum amount of water that a soil can hold after free drainage has ceased. It is the upper limit of available water. The surplus water is drained by large pores. The WHC is a key soil property because all chemical and biochemical processes require water. The capacity to store plant available water varies greatly among soils depending on their texture, depth, structure and humus content. The loamy/silty soils store the highest amounts of plant available water, whereas coarse sands store very little. Clay soils store considerable amounts of water, but a large portion is not available because it is tightly held in very small pores. Soil water is retained by adsorption and capillary forces, which are measured either in kiloPascals or by its related logarithmic pF value. Crops generally use 300–800 litres of water to produce 1 kg of dry matter. Some practical aspects of soil water are:

- Water is held mainly in medium-sized pores, and medium-textured soils hold the highest amounts.
- Only the free or loosely bound portion of water is available to plants. Water drains freely from pores with a diameter exceeding 60 μm. Thus, pores in the range 0.2–60 μm are important in retaining plant available water.
- Nutrient ions travel to the roots as part of the soil water.
- Roots can extract available water because of their suction forces.
- Maximum soil water storage against gravitational losses is at field capacity.
- Dry soils at the wilting point of crops have only non-available water left.
- The storage capacity within 150 cm depth varies from 40 to 120 mm of rain for most soils.

Soil air

Soil air is generally similar in composition to atmospheric air except that it has 7–10 times higher concentration of CO_2 than does the atmosphere. As a result of the respiration by roots and micro-organisms, the oxygen in the soil air may be consumed quickly and CO_2 produced, which is unfavourable for both root growth and functions. For most crops, the soil air should contain more than 10 percent oxygen but less than 3–5 percent CO_2. A continuous exchange with atmospheric air, termed soil aeration, is required in order to avoid a deficiency of oxygen. In cropped fields, the breaking of surface compaction can assist in this, but it must be done without destroying soil aggregates. In terms of air–water relations, the two extremes are represented by well-aerated sandy soils and the flooded-rice soils.

Soil physical properties and root growth

Crop growth requires that nutrients be present in soil in adequate amounts and in suitable forms for uptake. In addition, the nutrients must be supplied to the root surface at a sufficient rate throughout the growth of the crop so that the crop does not suffer from inadequate nutrient supply. This is particularly important during periods of rapid growth when nutrient demands are high. The physical nature of the soil affects the growth of an established plant through its influence on various factors such as aeration and moisture supply. In addition, such physical properties alter the resistance offered to root elongation and enlargement, proliferation and water uptake, which in turn affect plant nutrition. There are at least three important factors that determine the rate of root elongation. These are: turgor pressure within cells, constraint offered by the cell wall, and constraint offered by the surrounding medium. All of these are affected by the soil physical environment in the vicinity of the elongating root. The requirements of plant roots in soils are:

- deep rooting volume, ease of penetration and no restrictions on root growth;
- adequate available plant nutrients from soil reserves, external inputs or from N fixation;
- sufficient available water to support plants and soil life, for nutrient transformations and for nutrient transport to the roots;
- facility for the drainage of excess water from the rootzone to ensure the right air–water balance;
- good soil aeration to meet the oxygen requirements of roots and for the removal of surplus CO_2.

Root growth and the dynamics of water and air are largely dependent on pore space. Root growth occurs within continuous soil pores, within disturbed zones resulting from macro-organism activity and within the soil matrix itself. Pore size distribution is important for root penetration, water retention and aeration. In general, roots take the path of least resistance as they grow in soil. Root growth is reduced where the pore size is smaller than the root diameter because the plant must spend energy to deform the pore. The existence of sufficient continuous pores of adequate size is an important determinant of root growth. Most of the roots are 60 μm or more in diameter. The first-order laterals of cereals may range in size from 150 to 170 μm. In contrast, the root hairs are much smaller than 60 μm.

Bulk density is an indirect measure of pore space within a soil. The higher the bulk density, the more compact is the soil and the smaller is the pore space. In addition to absolute pore space, bulk density also affects the pore space distribution. Soil compaction decreases the number of large pores and, as these are the ones through which roots grow most easily, compaction can have an adverse effect on root growth.

The effect of bulk density may be altered considerably by changing the moisture content of the soil. As the pore space can be filled with either air or water and there is an inverse relationship between these two parameters, an increase in moisture content means a decrease in air-filled pores. In general, a decrease in soil moisture content reduces root growth even though more space is physically available to roots. Moreover, where the soil moisture content exceeds field capacity, this leads to poor aeration and root growth declines. Table 1 the effects of bulk density and aeration on plant growth. It shows that compaction of the soil under wet conditions can result in a marked decrease in root and top growth through a combination of mechanical impedance and aeration problems. There is a positive response to moisture in loose soil because the large pores drain easily and plant can suffer from a shortage of water. In contrast, adding water to the compact soil reduces root growth because of a lack of air in the soil pores caused by the displacement of air by water.

Table 1

The effect of moisture and of soil compaction on the growth of maize plants

Treatment	*Weight*	*Weight of tops*	*Weight of roots*	*Top:root of total plant*	*Ratio*
Compaction	Moisture		(g)		
Loose	wet	39.4	14.8	54.2	1:0.38
Loose	dry	27.5	9.3	36.8	1:0.34
Compact	wet	16.0	6.5	22.5	1:0.40
Compact	dry	20.1	11.3	31.4	1:0.56

Organic matter and soil fertility

The effect of SOM on soil fertility far exceeds its percentage share of the soil volume. Organic matter affects soil fertility and productivity in many ways:

- It promotes soil structure improvement by plant residues and humic substances leading to higher WHC, better soil aeration and protection of soil against erosion.
- It influences nutrient dynamics, particularly:
  - nutrient exchange, thus keeping the nutrients in available forms and protecting them against losses;
  - nutrient mobilisation from decomposed organic nutrient sources: N, P, S, Zn, etc.;
  - nutrient mobilisation from mineral reserves by complex formation or by changes in pH and redox potential;
  - immobilisation of nutrients on a short-term or long-term basis (reverse of mobilisation);
  - nutrient gain as a result of N fixation from the air.
- It influences promotion or retardation of growth through growth hormones.

Organic substances in the soil are important nutrient sources. Moreover, some substances can bring about the mobilisation of nutrients from soil mineral reserves by the production of organic acids, which dissolve minerals, or by chelating substances excreted by roots and/or by microbes. Chelates may bind Fe from iron phosphate and, thus, liberate phosphate anions. Organic matter may also have some negative effects, namely the short-term fixation of nutrients such as N, P and S into micro-organisms, which may create a transient deficiency particularly at wide ratios of C with these elements. The long-term fixation of these elements into stable humic substances appears to be a loss but it can be beneficial because of its positive effect on soil aggregation and, hence, on soil structure.

Rapid and far-reaching loss of SOM is an important factor in soil degradation. Many of the effects of organic matter are connected with the activity of soil life.

Soil organisms and soil fertility

Soil abounds in the following various types and forms of plant and animal life:

- animal life (fauna):
  - macrofauna (earthworms, termites, ants, grubs, slugs and snails, centipedes and millipedes),
  - microfauna (protozoa, nematodes and rotifers);

- plant life (flora):
  - macroflora (plant roots, and macro-algae),
  - microflora (bacteria, actinomycetes, fungi and algae).

Beyond the soil-forming activities of earthworms, termites and other large soil fauna, the multitude of different soil organisms contributes significantly to the soil physical and chemical conditions, especially in the transformation of organic matter and plant nutrients. The rate of transformation of most nutrients into available forms is controlled largely by microbial activity. Their huge number represents an enormous capacity for enzyme-based biochemical processes. A special case is N fixation by N-fixing free-living or symbiotic bacteria. Another case relates to the solubilisation of insoluble phosphates by several types of soil micro-organisms.

Soil micro-organisms have similar requirements of soil conditions for optimal activity in terms of air, moisture and pH, as do crops. In general, fungi are more active under acidic conditions, while bacteria prefer neutral–alkaline reaction. Any improvement in soil fertility for crops should also improve conditions for the activity of soil flora and fauna. Microbial activity not only determines soil fertility but it also depends on good soil fertility.

Soil Physico-chemical and Chemical Properties

Three important physico-chemical characteristics of soil fertility are: (i) soil reaction or pH; (ii) nutrient adsorption and exchange; and (iii) oxidation-reduction status or the redox potential.

Soil reaction

The reaction of a soil refers to its acidity or alkalinity. It is an important indicator of soil health. It can be easily measured and is usually expressed by the pH value. The term pH is derived from Latin *potentia Hydrogenii* and is the negative logarithm of the H^+ ion concentration. Because of the logarithmic scale used, in reality, the actual degree of acidity has enormous dimensions, e.g. the difference in acidity between pH 4 and 5 is tenfold. Thus, a soil of pH 5 is 10 times more acid than a soil of pH 6 and a soil of pH 9 is 10 times more alkaline than a soil of pH 8. The importance of soil pH is:

- the pH value indicates the degree of acidity or alkalinity of a medium, in this case soil;
- pH 7 is the neutral point, pH of 6.5–7.5 is generally called the neutral range;
- acid soils range from pH 3 to 6.5, alkaline soils from pH 7.5 to 10;

- most soils are in the pH range of 5–8, while the range for plant growth is within pH 3–10;
- the pH of a soil can be altered by amendments and nutrient management practices.

Soil pH is measured in soil/water suspensions. Where dilute calcium chloride solution is used instead of water, the data are lower by 0.5–1.0 of a unit. The pH value obtained is an average of the volume tested. In nature, there is a natural tendency towards soil acidification, the rate of which often increases under leaching, intensive cropping and persistent use of acid-forming fertilisers. Strong acidification leads to soil degradation. However, this can be overcome by the application of calcium carbonate or similar soil amendments. Unfavourable high pH values, as observed in alkali soils, can be decreased by amendment with materials such as gypsum, elemental S or iron pyrites.

Soil reaction is not a growth factor as such but it is a good indicator of several key determinants of growth factors, especially nutrient availability. Soil reaction greatly influences the availability of several plant nutrients. Phosphate is rendered less available in the strongly acidic upland soils. The availability of heavy metal nutrients (Cu, Fe, Mn and Zn) increases at lower pH, except for Mo. Although not a nutrient, Al becomes toxic below pH 4.5. Most plants grow well in the neutral to slightly acid range (pH 6–7) with the dominant cation Ca.

Plants are generally more sensitive to strong alkalinity, where the dominant cation is often Na, than to strong acidity where the dominant cation is H. The range of slight and moderate acidity can have special advantages in respect of nutrient mobilisation. Soils with very strong acidity (below pH 4.8) contain high levels of soluble Al. Almost no plants can survive below pH 3. The preference of plants for a certain pH range is often determined by aspects of nutrient requirement and efficiency and not because of the pH as such. Oats prefer a slightly acid range because of better Mn supply. Tea bushes benefit from an acid environment. This preference could also be caused by the adaptation of a plant species to a certain environment over time. However, this does not mean that the indicated crops cannot be grown outside the depicted pH range.

NUTRIENTS IN SOILS AND UPTAKE BY PLANTS

Soil Nutrient Sources

Many soils have vast reserves of plant nutrients but only a small portion of these nutrients becomes available to plants during a year or cropping season. However, all forms must change themselves to specific mineral ionic forms in order to be usable by plant roots. Thus, in order to become available to plants, nutrients must be solubilised or released from mineral sources and mineralised from organic sources including SOM.

Although nutrient mobilisation is a rather slow process, it increases sharply with temperature.

A temperature increase of 10 °C doubles the rate of chemical reactions. Consequently, the 20–30 °C higher temperature in tropical areas results in chemical transformations at 4–6 times higher the rate in temperate areas. About 1–3 percent of SOM is decomposed annually and this is a key determinant of N supply. If a fertile soil contains 8 000 kg N/ha in the organic matter, this corresponds to 160 kg of N transformed from organic N into ammonia, which may then be converted into nitrate. From this amount, crops may utilise about 50 percent, some is taken up by micro-organisms and some lost by leaching, denitrification and volatilisation.

As crop yields have increased over the years as a result of technological changes, few soils are able to supply the amounts of nutrients required to obtain higher yields without external inputs. An ideal soil is rich in mineral and organic sources of plant nutrients. In addition, it has the following characteristics:

- It has a strong capacity to mobilise nutrients from organic and inorganic sources.
- It stores both mobilised and added nutrients in forms that are available to plant roots, and protects them against losses.
- It is efficient in supplying all essential nutrients to plants according to their needs.

Practical importance of nutrients

A number of plant nutrients are of large-scale practical importance for successful crop production in many countries. Prominent among these are N, P, K, S, B and Zn. This means that their deficiencies are widespread and external applications are necessary to augment soil supplies for harvesting optimal crop yields while minimising the depletion of soil nutrient reserves.

N deficiency is widespread on almost all soils, especially where they are low in organic matter content and have a wide C:N ratio. Rare exceptions are soils with very high N-rich organic matter content during the first years of cropping, e.g. after clearing a forest. Widespread N deficiency is reflected in the fact that out of the 142 million tonnes of plant nutrients applied worldwide through mineral fertilisers, 85 million tonnes is N. In addition, substantial external N input is received through organic manures, recycling and BNF.

P deficiency was serious before the advent of mineral fertilisation because the native soil phosphate was strongly sorbed in very acid soils or precipitated as the insoluble calcium phosphate in alkaline soils. P deficiencies continue to be a major production constraint in many parts of the world. External input through mineral fertilisers alone was 33.6 million tonnes P_2O_5 in 2002. K deficiency is most strongly expressed in acid red and lateritic soils or on organic soils that have few K-bearing minerals. Soils rich in 2:1-type clays and those in arid or semi-arid areas are generally better supplied with K than soils in humid regions because of lower or no leaching losses in the former.

Ca supply is abundant in most neutral–alkaline soils and, hence, field-scale Ca deficiencies are rare. Where a Ca deficiency occurs, it is mainly in acid soils or because of insufficient uptake and transport of Ca within the plant. Mg deficiency can be widespread in acid soils as a consequence of low supply and leaching losses. S deficiency was of little practical importance decades ago because of considerable supply from the atmosphere, and widespread use of S-containing fertilisers such as single superphosphate (SSP) and ammonium sulphate. However, S deficiency has developed rapidly in recent years as the atmospheric inputs have declined and high-analysis S-free fertilisers have dominated the product pattern used. The problem has become significant and soil S deficiencies have been reported in more than 70 countries. About 9.5 million tonnes of S are currently applied as fertiliser worldwide.

Micronutrient deficiencies are common because of certain soil conditions and have developed at higher yield levels and on sensitive crops. On a global scale, the deficiencies of Zn and B are perhaps of greatest importance. Fe and Mn deficiencies frequently occur on calcareous soils or on coarse-textured soils with neutral or slightly alkaline reaction and rarely on acid soils. In certain areas, Mo deficiencies can even impede the establishment of legume pastures and lessen the potential gains from BNF. On strongly acid soils, there may even be problems of micronutrient toxicities.

Available Nutrients in Soils

Out of the total amount of nutrients in soils, more than 90 percent is bound in relatively insoluble compounds or is inaccessible within large particles and, therefore, is unavailable for crop use. Only a very small proportion is available to plants at any given point of time. To assess the nutrient supply to crops, it is important to know the amount of available soil nutrients either actually present or likely to be accessible to the plant during a cropping season. All available nutrients must reach the rootzone in ionic forms that plant roots can take up. In order for plants to acquire available nutrients, plant roots must intercept them in the soil or they must move to the root either with the water stream or down a chemical concentration gradient. Moreover, available nutrients in soils are not a specific chemical entity or a homogeneous pool, but consist of three fractions. In terms of decreasing availability, these are:

- nutrients in the soil solution;
- nutrients adsorbed onto the exchange complex;
- nutrients bound in water-insoluble forms but easily mobilisable nutrient sources.

While the first two fractions are easily available and can be determined by fairly accurate and precise methods, the third fraction comprises a range of substances with varying availability and, therefore, is difficult to assess.

Nutrients in the soil solution

The soil solution is the substrate from which roots take up nutrients. It is comparable with the nutrient solution in hydroponics. Soil solution means soil water containing small amounts of dissolved salts (cations/anions) and some organic substances that is mainly held in medium to small pores. The concentration of these nutrients is very different and varies considerably in time.

The solution of fertile soils may contain 0.02–0.1 percent salts in wet soils but a higher concentration in dry soils. In a neutral soil, the dominant nutrient is generally nitrate (30–50 percent), followed by Ca (20–30 percent), and Mg, K and sulphate-S. Ammonium is less than 2 percent and phosphate-P is considerably less than 0.1 percent. In saline and alkali soils, there are large concentrations of Na, chloride and sulphates. Compared with the daily nutrient requirement of high-yielding crops, the amounts of nutrients in the soil solution are very low and can only meet plant needs for a short period. This is especially true for phosphate and micronutrient cations. At high rates of nutrient uptake, the soil solution soon becomes depleted where it is not replenished from other fractions of available nutrients and unavailable forms, just as well-water is replenished by the groundwater resources as the water is drawn from the well. Adsorbed forms and other potentially available nutrient forms continuously replenish the soil solution, which represents the intensity factor. Adequate fertilisation ensures such replenishment. Where this cannot take place, nutrient deficiencies are very likely to occur.

Nutrients on the exchange complex

The fraction of exchangeable nutrients is much larger than their amount in the soil solution. In fertile soils of neutral soil reaction, about 75 percent of the adsorbed cations are of Ca and Mg and only 3–5 percent are of K. The adsorbed nutrients and anions, especially phosphate, are protected to a considerable extent against leaching loss and yet are easily available. An undesirable reality is that large amounts of nitrate are not adsorbed at all and, hence, can be easily lost through leaching.

Nutrients on the exchange complex must first be desorbed, exchanged or released into the surrounding soil solution before they can be taken up by plants. These replenish the soil solution. There is a steady exchange between nutrients on the exchange complex and those in the soil solution. A cation exchanges places with a cation and an anion can only exchange with another anion. Desorption dominates where the solution is diluted by nutrient uptake or addition of water, whereas adsorption dominates after input of water-soluble nutrient sources or with increasing dryness. Plant roots contribute to the release of adsorbed cations by the production of hydrogen ions (H+), which may replace other exchangeable cations.

For some nutrients, the exchangeable fraction is a fairly good indicator of the total available pool of a nutrient in many soils. This is especially so in the case of K and Mg except for soils that can release significant amounts of non-exchangeable K during the crop season. The inaccessible cations are those that are within the close-packed interior of clay mineral layers. These are regarded as fixed and mostly non-available. This is especially the case with nutrients such as K^+ and NH_4^+.

K fixation in soils with certain clay minerals can result in severe K deficiencies. However, recent studies of several soils, specially the illite-dominant alluvial soils of India, indicate a very substantial contribution of non-exchangeable or "fixed" fraction of K to K uptake by crops. This calls for a change in thinking regarding the practical importance of non-exchangeable K for crop nutrition.

Moderately available nutrients

These nutrients are bound within different insoluble mineral and organic sources but are released during the cropping season. They can be easily mobilised by dissolving agents produced by micro-organisms or by plant roots. Some phosphate may be mobilised by organic acids and by mycorrhisae while some micronutrient cations by organic complexing agents known as chelates. The non-exchangeable K referred to above can also be considered in this category of moderately available nutrients.

It is difficult to distinguish this group from the much larger and partly similar pool of non-available nutrient sources as they are in a continuum. For example, typical P-containing compounds are calcium, aluminium and iron phosphates, but whether they belong to the moderately available nutrients depends on several factors. Moderately available P comes from freshly precipitated surface layers of amorphous material of small particle size, which facilitates their dissolution by dilute mineral and organic acids or by complexation. Examples of non-available forms of P are the occluded forms and tricalcium phosphates in alkaline soils.

The same phosphates can become moderately available in a strongly acid soil. Insoluble iron phosphates can become available in reduced paddy soils where the ferric form becomes reduced to the soluble ferrous form of Fe. Thus, the borderline between available and non-available nutrients in chemical compounds is arbitrary. Therefore, the amount of nutrients released into easily available forms during a cropping season is difficult to assess very accurately via practical approaches.

DYNAMICS OF PLANT NUTRIENTS IN SOILS

The content of available nutrients and their degree of availability and accessibility is not a static condition for all situations but ever-changing and very dynamic because of the various inorganic and biochemical processes that take place in soils. These depend on temperature, water content, soil reaction, nutrient uptake, input and losses, etc. Most forms of a nutrient are in a dynamic equilibrium. External applications only cause temporary changes in the relation between different fractions, but the basic nature of the equilibrium remains intact over time.

An increasing water content, a stronger sorption of divalent cations and an increase in the mobilisation rate. With increasing dryness, the soil solution becomes more concentrated and contains relatively more divalent cations, but, most important, with dryness there is an increased immobilisation of nutrients into only moderately available forms.

A decrease in pH from the neutral range results in a smaller proportion of exchangeable Ca and Mg. In the case of phosphate, there is initially a greater mobilisation of calcium phosphate, but later a strong immobilisation or even fixation into aluminium and iron phosphates. The availability of some micronutrients, especially of Fe, Mn and Zn, is increased strongly, and can even reach toxic levels.

An increase in pH by liming can reverse the situation. Nutrient uptake by plants, biological activity of soil organisms and external nutrient input can result in large or small fluctuations among the nutrient fractions, resulting in an ever-changing soil fertility status. To a certain extent, this can and should be controlled by appropriate management practices. As a result of nutrient transformation and dynamics, when a nutrient ion reaches the root surface, the plant cannot distinguish whether this nutrient has come from soil reserves, mineral fertilisers or organic manures. In all probability it does not make any difference to crop nutrition.

DYNAMICS OF MAJOR NUTRIENTS

Nitrogen Dynamics

The dynamics of N in soils are quite complex. These are depicted in many ways in the scientific literature. Many factors affect the level of plant available inorganic N. Soil N is primarily in the organic fraction. The N in the organic matter came initially from the atmosphere via plants and micro-organisms that have since decomposed and left resistant and semi-resistant organic compounds in the soil during development. As the bulk of the organic matter is in the upper horizons, most of the soil N is also in the topsoil.

Inorganic ionic forms of N absorbed by roots usually constitute less than 5 percent of total soil N. In normal cropped soils, where ammonium is added through fertilisers or released from organic matter/organic manures/crop residues by mineralisation, it is usually nitrified rapidly to nitrate. N added in the amide form (NH_2) as in urea is first hydrolysed to NH_4^+ with the help of urease enzyme. It can then be absorbed by roots as such or converted to nitrate and then absorbed. Where urea is left on the soil surface, particularly on alkaline soils, some of it can be lost through ammonia volatilisation. Such ammonia can return to the soil with rain.

The ratio of NH_4^+ to NO_3^+ in soil depends on the presence of satisfactory conditions for nitrification, which is inhibited by low soil pH and anaerobic conditions. The nitrate thus formed can be absorbed by roots, immobilised by micro-organisms and become part of organic N, move down with percolating waters and leach out of the rootzone or be denitrified under anaerobic conditions. In cases where groundwaters are pumped for irrigation, the leached nitrate can re-enter the soil with irrigation water. Certain 2:1-type clays such as vermiculite and illite can fix ammonium ions in a non-exchangeable form. Gaseous forms of N include dinitrogen (N_2) from the atmosphere or denitrification or nitrous oxides from denitrification. N in the soil atmosphere can only be used by symbiotic N-fixing bacteria such as *Rhizobium* or non-symbiotic N-fixing bacteria such as *Azotobacter, Azospirillum, Cyanobacteria* and *Clostridium*. Several nitrogen gases that escape from the soil after denitrification or volatilisation return to the soil with rain. The nitrogen gas itself can return to the soil through biologically or industrially fixed N.

Phosphate dynamics

Chemically, P is one of the most reactive plant nutrient. Thermodynamic principles dictate that P compounds will tend to transform to less soluble and increasingly stable forms with the passage of time. Hence, P is one of the most unavailable and immobile nutrient elements. One of the indicators of this is that barely 15–20 percent of the P added through fertiliser is recovered by the crop. It exists in the soil in a variety of forms. The dynamics of phosphate in soil present special problems because of the low solubility of most P compounds.

P added through soluble fertilisers first enters the soil solution, but much of it is converted into adsorbed P within a few hours. Very little of the added P stays in the soil solution except in very sandy soils or soils lacking in the P-adsorbing agents. P added through PR is first solubilised by soil acids, after which it participates in various dynamic reactions as in case of soluble P. Some important aspects of P dynamics are:

- Soil solution P: It is present in very small amounts and ranges from 0.01 to 0.50 mg/litre. In comparison, the concentrations of nutrients such as Ca, Mg and K are of the order of 400, 60 and 40 mg/litre, respectively. The relative distribution of anionic forms of P is dependent on soil pH. In the common pH range for soils, the dominant ionic form is $H_2PO_4^-$, which is also the most common form absorbed by plants. As the pH increases, the relative proportion of $H_2PO_4^-$ decreases while that of HPO_4^{2-} and PO_4^{3-} increases. In addition to ionic P in solution, some solution P may be present as soluble organic compounds, particularly in soils containing appreciable quantities of organic matter.

- *Adsorbed P*: In acid soils, the reactive phosphate ion is adsorbed onto the surfaces of iron and aluminium hydrous oxides, various clay minerals and aluminium–organic matter complexes. In neutral and alkaline soil, inorganic P may be adsorbed onto the microsurface of calcium and magnesium carbonates, iron and aluminium hydrous oxides, various clay minerals and calcium–organic matter complexes. Adsorbed P is a major source of P extracted by reagents used to estimate available P. It is in a dynamic equilibrium with solution P and replenishment as the P from solution is used up.
- *Mineral P:* These are mainly minerals of P combined with Ca, Al and Fe. In soils above pH 7, calcium and magnesium phosphate are dominant, while iron and aluminium phosphates are the dominant forms in acid soils. The amorphous forms can contribute to plant nutrition, but the crystalline forms are more stable and less reactive.
- *Organic P:* One-half or more of the total soil P may be present as organic P, the amount depending on the content and composition of organic matter. The major P-bearing organic compounds in soil are inositol phosphate, phospholipids, nucleic acids and others such as esters and proteins. The net release of this P to plants depends on the balance between mineralisation and mobilisation.
- *Available P:* The replenishment of P into the soil solution following P uptake by plant roots is dependent on the quantity of P in the adsorbed and sparingly soluble mineral phase, as well as inorganic P in plant residues (Figure 2).

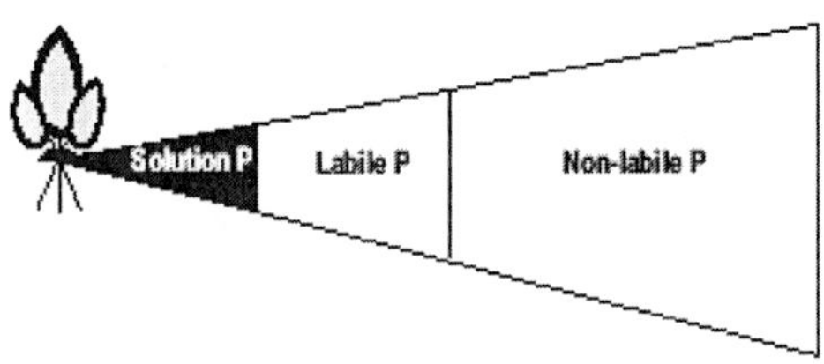

Figure 2: Schematic representation of the three important soil P fractions for plant nutrition

Together, these constitute the reservoir of "readily available P" for soil solution replenishment. This is also known as labile P and is usually defined as the quantity of P that is in equilibrium with the soil solution during the life of the plant. The ability of a soil to maintain its P concentration in solution as plant roots absorb the P, or as P is added by fertilisation, is known as the P-buffering capacity. The higher the buffering capacity is, the larger is the proportion of P in the solid phase relative to the solution phase.

Potassium dynamics

Among cations, K^+ is absorbed by plants in the largest amount. The four important forms of K in soil are: (i) mineral K; (ii) non-exchangeable K, or K fixed in between clay plates; (iii) exchangeable K; and (iv) K present in the soil solution. The abundance of different K fractions is usually as follows: 90–98 percent of total K is in mineral form, which is relatively inaccessible to a growing crop; 1–10 percent of total K is in the non-exchangeable form, which is slowly available; and 1–2 percent of total K is in the exchangeable and water-soluble forms. Of this 1–2 percent, about 90 percent may be exchangeable and 10 percent in the soil solution. All these are in a dynamic equilibrium.

K dynamics are determined by the rate of K exchange from the clay and organic matter surfaces and the rate of release from soil minerals. Except for sandy soils, K+ is stored on the surface of negatively charged clay minerals. This easily replaceable supply provides the soil solution with additional K when the soil-solution K concentration decreases as a result of crop uptake. Most traditional soil test measurements of available K include the exchangeable and the water-soluble fractions. There is increasing evidence that a part of the nonexchangeable fraction or fixed K should also be included in soil test measurements for making meaningful K fertiliser recommendations.

The K in common fertilisers is water soluble. On addition to the soil, the fertiliser dissolves in the soil water and dissociates into the cation (K^+) and the anion (Cl^- or SO_4^{2}). The cation K^+ is largely held on to the exchange complex as an exchangeable cation and a small amount is present in a freely mobile form in the soil solution. Thus, the K added to soils can be transformed into three fractions apart from any incorporation into the organic matter.

In very sandy soils, particularly under high rainfall, K can also leach out of the rootsone. Similarly, a significant amount of available K can be added to the soil through crop residues, leaf fall and irrigation water in specific cases. Much of this K is soluble and can again be adsorbed on the exchange complex.

Calcium dynamics

The Ca content of different soils varies greatly, depending on the minerals from which the soils originate and on the degree of weathering. Ca content is lowest in acid peat soils and in highly weathered soils of the humid tropics where, owing to high acidity and heavy leaching losses, as little as 0.1–0.3 percent Ca may be left. On the other hand, calcareous soils sometimes contain more than 25 percent Ca, mostly in the form of calcium carbonate. In general terms, Ca in the soil occurs in the following forms:

- Ca-bearing minerals such as calcium aluminium silicates (e.g. plagioclase), calcium carbonates (calcite and dolomite), calcium phosphates (apatite), and calcium sulphate (gypsum);
- exchangeable Ca, adsorbed on negatively charged surfaces of organic and inorganic soil colloids as part of the CEC of the soil;.
- water-soluble Ca.

As is the case with other cations, the different forms of Ca in the soil are in a dynamic equilibrium. Plant roots take up Ca from the soil solution. When the Ca concentration in the solution decreases, it is replenished by the exchangeable Ca. An exchange complex dominated by Ca and adequately provided with Mg and K is a favourable precondition for good crop yields. In humid temperate regions, the ratio of the nutrients is considered well-balanced when about 65 percent of CEC is saturated by Ca, 10 percent by Mg, 5 percent by K and the remaining 20 percent by others (H, Na, etc.).

Magnesium dynamics

From the viewpoint of plant nutrition, the Mg fractions in the soil can be considered in a similar manner to those of K. Mg on the exchange complex and in the soil solution is most important for plant nutrition. Mg saturation of the CEC is usually lower than that of Ca and higher than that of K. The major forms of Mg are:

- non-exchangeable Mg;
- exchangeable Mg;
- water-soluble Mg.

Non-exchangeable Mg is contained in the primary minerals such as hornblende or biotite, as well as in secondary clay minerals such as vermiculite, which hold Mg in their interlayer spaces. Non-exchangeable Mg is in equilibrium with the exchangeable Mg, which in turn replenishes the water-soluble or solution Mg. If the exchangeable Mg is depleted, as under exhaustive cropping, plants will utilise Mg from originally non-exchangeable sites at the clay minerals. However, its rate of release is too slow for optimal plant growth.

Sulphur dynamics

The S content of soils is usually lower than that of Ca or Mg. It is in the range of milligrams per kilogram or parts per million rather than percent. In the soil, S occurs in organic and inorganic forms. Tropical soils generally contain less S than do soils in temperate regions because of their lower organic matter content and its rapid rate of decomposition. In a survey of S in the tropics, an average value of 106 mg/kg S for a wide range of tropical upland soils has been reported, which is well below the 200–500 mg/kg S reported for non-leached temperate soils in the United States of America.

In the soil, inorganic S exists mostly as sulphate. This is either readily soluble or adsorbed on soil colloids. In calcareous soils, sulphate may also be present as cocrystallised impurity with calcium carbonate. Some soils have gypsic horizons that are enriched with the sulphates of Ca and Mg. As plant roots take up S from the soil solution in the form of sulphate, only the soluble and the adsorbed SO_4^{2-} are readily available. In many soils, these fractions represent not more than 10 percent of the total S.

S present in SOM or S added through organic manures becomes available for plant use only after conversion to the sulphate form through mineralisation. Conditions most favourable for the mineralisation of organic S are: high temperatures; a soil moisture status that is about 60 percent of field capacity; conditions favourable for high microbial population; and a minimum S content of 0.15 percent in the organic matter.

When water-soluble sulphate fertilisers such as ammonium sulphate (AS) and single supersulphate are added, the sulphate enters soil water. From there, it can either be moderately adsorbed or immobilised in soil organic water after absorption by micro-organisms. A small portion remains in the form of sulphate ion in the soil solution. Where S is added in elemental or sulphide forms, as in pyrites, these undergo oxidation in the soil to furnish sulphate ions for plant use. This transformation is affected by soil pH, moisture status, aeration and particle size of the S carrier.

ASSESSMENT OF NUTRIENT STATUS OF SOILS AND PLANTS

The evaluation or assessment of soil fertility is perhaps the most basic decisionmaking tool for balanced and efficient nutrient management. It consists of estimating the available nutrient status of a soil for crop production. A correct assessment of the available nutrient status before planting a crop helps in taking appropriate measures for ensuring adequate nutrient supply for a good crop over and above the amounts that the soil can furnish. The techniques used include soil testing and plant analysis, the latter including related tools such as total analysis of the selected plant part, tissue testing, crop logging and the diagnosis and recommendation integrated system (DRIS). The objective of all these techniques is to assess the available nutrient status of soils and plants so that corrective measures can be taken to ensure optimal plant nutrition and minimum depletion of soil fertility.

Soil Testing

Soil testing is the most widely used research tool for making balanced and profit maximising fertiliser recommendations, particularly for field crops. Soil testing can be defined as an acceptably accurate and rapid soil chemical analysis for assessing available nutrient status for making fertiliser recommendations. Soil testing as a diagnostic tool is useful only when the interpretation of test results is based on correlation with crop response and economic considerations to arrive at practically usable fertiliser recommendations for a given soil–crop situation.

The amount of a nutrient estimated as available through soil testing need not be a quantitative measurement of the total available pool of a nutrient but a proportion of it that is correlated significantly to crop response. Soil testing does not measure soil fertility as a single entity but the available status of each nutrient of interest is to be determined. Based on a high degree of correlation between the soil test value of a nutrient and the crop response to its application, the probability of a response to nutrient input can be predicted. This serves as a basis for making practical fertiliser recommendations, which should be adjusted for nutrient additions expected to be made through BNF and organic manures. Soil testing has to be done for each individual field and for each nutrient of interest. It may be repeated every 3–4 years. The major steps in practical soil testing for a relatively uniform field of up to 1 ha are:

- representative soil sampling of the fields;
- proper identification and labelling of the sample;
- preparation of soil sample;
- extraction of available nutrients by an appropriate laboratory method;
- chemical determination of extracted nutrients;
- interpretation of soil analysis data – soil test crop response correlation is the key issue.

The usefulness of soil testing depends on a number of factors, such as representative and correct soil sampling, analysis of the sample using a validated procedure, and correct interpretation of the analytical data for making practical recommendations. A sound soil testing programme requires an enormous amount of background research on a continuing basis to cater to changing needs such as the development of new crop varieties, better products and agronomic practices. Such research also helps to determine:

- the chemical forms of available nutrients in soils and their mobility;
- the most suitable extractants for accurately and rapidly measuring such forms;

- the general health and productivity of the soils for various crops;
- norms for field soil sampling and sample processing techniques;
- the response of crops to rates and methods of fertiliser application;
- the effect of season on nutrient availability;
- interactions of a nutrient with moisture and other nutrients.

Plant Diagnosis

The nutrient status of plants can by assessed on a qualitative basis by visual observation and, more accurately, on a quantitative basis by analysing the mineral composition of specific parts of growing plants.

Visual plant diagnosis

A healthy dark-green colour of the leaf is a common indicator of good nutrient supply and plant health. The degree of "greenness" can be specified in exact terms for each crop using Munsell's Plant Colour Chart or other such charts. Any change to light green or a yellowish colour generally suggests a nutrient deficiency where other factors are not responsible such as cold weather, plant diseases, air pollution, etc.

A deviation from the normal green colour is easily detected. However, it is by no means always caused by N deficiency, as usually assumed. For example, it is a common but questionable practice to always relate light-green late foliar discolouring with N deficiency, but it can often be caused by other deficiencies. Even where the colour is more or less "satisfactory", there may be a latent deficiency that is often difficult to establish from visual observation but can still cause yield reduction. It usually requires chemical plant analysis for conformation. Fully developed deficiency symptoms can be a useful means for detecting nutrient deficiencies. However, they are only reliable where a single nutrient and no other factors are limiting. However, some general guidelines for the appearance of nutrient deficiencies in cereals are:

- Deficiencies indicated by symptoms appearing first on older leaves:
  - chlorosis starting from leaf tips, later leaves turn yellowish-brown: N;
  - reddish discoloration on green leaves or stalks: P;
  - leaves with brown necrotic margins, wilted appearance of plant: K;
  - stripe chlorosis, mainly between veins, while veins remain green: Mg;
  - spot necrosis: greyish-brown stripe-form spots in oats: Mn;
  - dark-brown spots in oats and barley, whitish spots in rye and wheat: Mn.

- Deficiencies indicated by symptoms appearing first on younger leaves:
  - completely yellowish-green leaves with yellowish veins: S;
  - yellow or pale yellow to white leaves with green veins: Fe;
  - youngest leaf with white, withered and twisted tip (oats and barley). Cu;
  - yellowish leaves with brownish spots (part of acidity syndrome): Ca.

Chlorosis refers to a condition in which the leaves appear with a light greenyellowish tinge, but the tissue is still largely intact. Necrosis means a brownish dark colour with irreversibly destroyed tissue. The easiest way of visually diagnosing nutrient deficiency symptoms is their identification with good-quality colour photographs of the specific crops. Even with these, farmers are advised to seek professional help and plant analysis as needed before taking corrective measures.

Plant analysis

The nutrient concentration of growing plants provides reliable information on their nutritional status in most cases, expect in the case of Fe. It reflects the current state of nutrient supply and permits conclusions as to whether a supplementary nutrient application is required. Plant analysis generally provides more current plant-based information than soil testing but it is more costly and requires greater efforts in sampling, sample handling and analysis. Ideally, both tools should be used as they complement each other. The key features of plant analysis are:

- *Sampling:* Representative sampling should be done of specific plant parts at a growth stage that is most closely associated with critical values as provided by research data. Sampling criteria and the procedure for individual samples is similar to that for soil testing in that it should be representative of the field. The composite sample should be about 200–500 g fresh weight.
- *Sample preparation:* The collected sample should be washed as soon as possible first with clean water and then with distilled water. It should be air dried followed by oven drying at 70 °C. Finally, it has to be carefully ground, avoiding contamination, and the powder mixed well.
- *Analysis:* After dry or wet ashing and complete dissolution, the determination of nutrients by standard analytical methods is carried out. The results are expressed as a concentration on a dry-matter basis.
- *Interpretation:* Interpretation of plant analysis data is usually based on the total concentrations of nutrients in the dry matter of leaves or other suitable plant parts, which are compared with standard values of "critical nutrient concentrations" and

grouped into supply classes. This will determine whether immediate action such as foliar spraying is needed to correct a deficiency. Conclusions can also be drawn on whether the amount of fertiliser applied at sowing time was sufficient or should be increased for the next crop. Where the concentration is in the toxicity range, special countermeasures are required but no application is needed.

For some nutrients, such as Ca and Fe, the "active" nutrient content of plants should be considered because immobilisation can make the total concentrations misleading. The nutrient concentrations of green material or of plant sap can be used as a suitable basis for interpretation in some situations. Because of many interactions between nutrients and other inputs, more sophisticated indicators than just individual concentrations have been suggested, such as simple or complex nutrient ratios, e.g. the DRIS method.

Critical values

The critical level is that level of concentration of a nutrient in the plant that is likely to result in 90 percent of the maximum yields. The plant nutrient concentrations required depend on the cultivar and expected yield level. Standard tables of plant analysis interpretation are based on requirement levels for very high yields or on those for the more practical and realistic medium to high yields corresponding to critical values of 90 percent. The main advantage of critical values, once properly established, is their wide applicability for the same crop. Their disadvantage is that they only provide a yes or no type of information and do not cover the entire range over which nutrient supplies need to be managed.

In most cases, the correspond to 90 percent of maximum yield but in some cases to maximum yield. These are approximations compiled from various sources. Specific situations require further refinement. For example, critical concentrations in the case of oil-palm are different for young palms and for older palms. A selection of critical plant nutrient concentrations for many crops has been compiled by the International Fertiliser Industry Association among others.

IMPACT OF SOIL FERTILITY ON CROP PRODUCTIVITY

The basic aim of sound soil fertility management is to enhance crop productivity, to sustain it, and to keep the soils in good health – physically, chemically and biologically.

Improving Soil Fertility

The high yield levels obtained are a result of suitable crop growth conditions, optimal and balanced nutrient management and adoption of best management practices. When taken year after year, such steps lead to an improvement in soil fertility, which makes the production of high yields sustainable. After more than a century of mineral fertilisation, along with organic supplements available to the farmers, there has been a considerable increase in organic and mineral contents of major nutrients in the soil in many parts of the world. At the same time, in large areas, soils continue to be mined of their nutrient reserves and are becoming depleted.

Such soils are losing their ability to sustain high levels of crop productivity and safeguard food security. Because of improved soil fertility, cereal yields on many highly productive soils remains high for some time even where left unfertilised. They may remain at 8–10 tonnes/ha in the first year, decrease to 5–7 tonnes/ha after some years, and further to about 3–4 tonnes/ha. Finally, after a longer unknown period, they can probably decline to 1–3 tonnes/ha. Wherever high yields are to be obtained on a sustained basis, the crop requires access to adequate amounts of all essential plant nutrients. Wherever the fertility of a soil is unable to furnish such amounts, soil fertility has to be improved through external additions of required plant nutrients. This is best accomplished through INM.

Degradation and Improvement of Soil Fertility

Soil degradation leads to a deterioration in soil quality, resulting in yield decline. Soil degradation lowers the actual or potential soil productivity in different ways:

- loss of the fertile topsoil components through erosion by water and wind;
- physical degradation;
- chemical and biological degradation, e.g. decrease in organic matter and soil bioactivity, loss of nutrients through various routes, soil acidification or salinisation with their accompanying problems of nutrient deficiencies, toxicities and imbalances.

Soil degradation is widespread in many parts of the world. The basic causes of soil degradation are the result of human activities such as deforestation, overgrazing and poor soil management. Factors that cause soil degradation are interrelated. About 1 200 million ha worldwide are considered to be affected by soil degradation, mostly by erosion. It has been estimated that human-induced soil degradation has affected 46 million ha in Africa and 15 million ha in Asia. Out of these, 25 percent of such soils in Africa and 67 percent in Asia are moderately to severely affected.

Degradation of soil fertility

Soil fertility is not a stable property but a dynamic one. There are widespread problems of soil fertility degradation under many cropping systems even on soils with good initial soil fertility. The result of such a decline is a reduced nutrient supply, which reduces crop yields.

From plant nutrition considerations, chemical degradation of the soil, particularly its fertility status, is of greatest concern. Losses of nutrients from soil can be caused by soil erosion, leaching, crop removal or in the form of gases. Nutrient removal by crop products compared with external nutrient inputs can be similar, higher or lower. Negative nutrient balances result where nutrient removals exceed nutrient additions. These are a cause of soil fertility depletion or nutrient mining. Positive nutrient balances indicate a buildup or improvement in soil fertility.

Nutrient mining or depletion is a widespread problem in low- and mediuminput agriculture. This is a major threat to productive sustainable farming. It is accelerated by imbalanced fertilisation. Nutrient mining can cause the exhaustion of any nutrient required in moderate to large amounts. It can be particularly severe in the case of N, P, K and S depending on soil nutrient reserves and the amounts replenished. A negative balance can be acceptable for a short period, but, where prolonged, it will lead to soil deterioration. It is expensive to improve depleted soils. Experience from Africa shows that, on poor soils, 33 percent of the total soil N may be lost within 10 years and 33 percent of P within 20 years, even at grain yields of 2 tonnes/ha. In comparison, K losses are relatively smaller and those of Zn are very small. Several countries have a negative nutrient balance of more than 60 kg of total nutrients annually. Negative nutrient balances are quite common also in many countries.

REFERENCES

Budelman, A. & Defoer, T. 2000. *Managing soil fertility in the tropics - PLAR and resource flow analysis in practice. Case studies from Benin, Ethiopia, Kenya, Mali and Tanzania.* Amsterdam, The Netherlands, Royal Tropical Institute.

Eyasu, E. 1998. *Is soil fertility declining? Perspectives on environmental change in southern Ethiopia.* Managing Africa's Soils: 2. London, Drylands Programme IIED.

FAO. 1980. *Improving soil fertility through organic recycling - comparative analyses of organic manures,* by R.V. Misra & P.R. Hesse. Project field document No. 24. Rome.

Hartemink, A.E. 2001. *Soil fertility in the tropics with case studies on plantation crops.* Reading, UK, The University of Reading.

Hilhorst, T. & Muchena, F.M. 2000. *Nutrients on the move - Soil fertility dynamics in African farming systems.* London, International Institute for Environment and Development (IIED).

5

Plant Propagation

Most ornamental plants in Florida can be propagated easily by home gardeners. By doing so, the gardener increases the number of plants with desirable characteristics and decreases landscaping costs. Plants can be propagated by asexual or sexual means. Sexual propagation involves starting plants from seed, while asexual propagation refers to multiplication of plants from vegetative plant parts such as shoots, roots, and leaves, or specialised organs such as bulbs and corms. Budding and grafting are also methods of vegetative propagation, but will not be addressed in this publication.

ASEXUAL PROPAGATION

The most important reason for asexual propagation is to grow plants with the same characteristics as the parent plant. Asexual propagation is the only practical means of reproduction when a plant does not produce viable seed, or seeds are difficult to germinate. A group of plants originating from a single plant and reproduced by vegetative means is called a clone. For example, the original Drake elm came from a seed, but has since been increased by vegetative propagation to maintain its desirable characteristics.

Cuttings

The most common method to propagate plants asexually is from cuttings. Cuttings can be made from stems, roots, leaves, or combinations of plant parts such as stems with leaves (Figure 1). Cuttings should be taken from healthy plants with desirable characteristics, and placed in a warm, humid environment to hasten root development and prevent them from drying.

m cutting *Subterminal stem cutting* *Leaf-bud cutting*

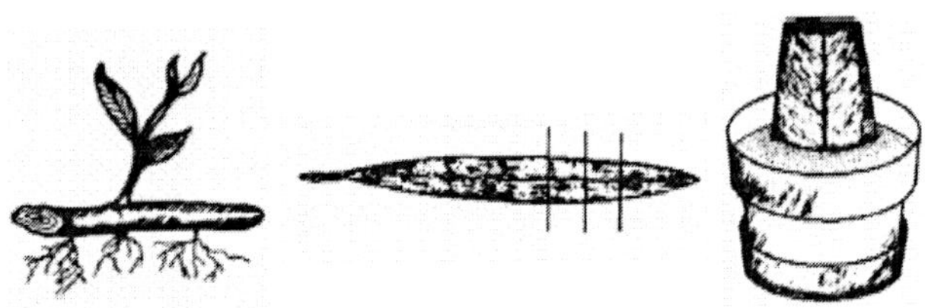

Root cutting *Leaf cutting (Ssansevieria)*

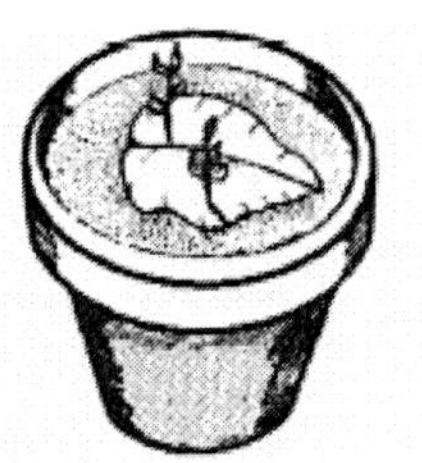

Leaf cutting (Begonia)

Figure 1. Types of Cuttings

Stem cuttings

Stem cuttings can be taken at different stages of vegetative maturity and may consist of just the growing tip of a plant or subterminal stem sections. Some plants root better from softwood cuttings, while others should be propagated from semi-hardwood or hardwood cuttings. Softwood and semi-hardwood cuttings are from the current season's growth, and hardwood cuttings (seldomly taken in Florida) are from the previous season's growth. Softwood cuttings are generally taken from plants in spring or early summer during a growth flush when the tissue is relatively soft and succulent. Semi-hardwood cuttings are taken after a growth flush has matured. Stems of semi-hardwood cuttings will usually "snap" like green beans when broken. Many Florida plants root best as semi-hardwood cuttings.

Stem cuttings are removed using a clean, sharp knife or pruner. Cuttings 4 to 6 inches (10 to 15 cm) in length are appropriate for most plants. Leaves are removed from the bottom 1 inch (2.5 cm) of stem cuttings, and then the cuttings are stuck upright in a propagation medium. Insert the cuttings just deep enough—usually ½ to 1 inch (1.2 to 2.5 cm)—into the propagation medium to hold them upright. A mixture of equal volumes of peat moss and coarse perlite is a suitable rooting medium for most plants, but combinations of other materials such as shredded sphagnum, vermiculite, and sand have also proven satisfactory. The medium should drain freely and be free of disease organisms and weed seed. Packaged media can be purchased, or small quantities can be sterilised by placing a 2-inch (5 cm) layer of moist medium on a tray in an oven at 220°F (104°C) for 1 hour. The odor from heated moist media may be offensive.

Root-promoting chemicals (hormones) can be applied to the basal 1/2 inch (1.2 cm) of cuttings before sticking them in a medium to enhance rooting of some plants. Root promoting chemicals are primarily composed of auxins: IBA and/or NAA. Although it is possible to obtain these chemicals and prepare your own, it is more practical to purchase the commercially prepared talc formulations. These commercial preparations are available at most garden centers in various concentrations, suited for easy-, moderate, or difficult-to-root plants. Some talc formulations of auxins also contain a fungicide to aid in preventing disease during rooting.

Leaf cuttings

Leaf cuttings may be comprised of only the leaf blade or the leaf blade and petiole. Begonias and sansevierias are commonly propagated by leaf cuttings. Leaf cuttings of some plants, such as the Rex begonia, are wounded by cutting the underside of the main veins before placing the leaf surface flat and in firm contact with the propagation medium. Sometimes it is helpful to pin these leaves to the moist medium with small stakes or toothpicks. Leaf cuttings of many plants can be stuck upright in the propagation medium. When subterminal sections of leaves are used, make sure the basal end of the cutting is inserted into the propagation medium. Roots and new shoots will start at the base of the leaf or at points where the veins were cut.

Leaf-bud cuttings

Leaf-bud cuttings include the leaf blade, the petiole, and a ½- to 1-inch segment of the stem. Axillary buds located at the union of the petiole and stem produce new shoots under warm, humid conditions. This method is often used for plants in short supply that have long internodes. Every node on the stem can be a cutting.

Root cuttings

Root cuttings are usually taken from young plants in early spring or late winter, before they start growing. Healthy roots have ample food (carbohydrates) stored to support shoot development at this time. Root cuttings are typically 2 to 7 inches (5 to 18 cm) in length depending upon root diameter. Large roots can be cut shorter than small roots and still have an adequate food supply for root and shoot initiation and growth. Small, delicate root cuttings (1/8 to ¼ inch or 3.2 to 6.4 mm in diameter) should be positioned horizontally in the propagation medium and covered with ½ inch of medium.

Larger root cuttings (1/4 to ½ inches or 6.4 to 12.8 mm in diameter) can be planted vertically with the end of the cutting originally nearest the plant crown positioned upward. Optimum temperatures for most root cuttings range from 55°F to 65°F (13°C to 18°C). Root cuttings may be transplanted after shoots have emerged and sufficient new secondary roots have developed. The principal disadvantage of this method is the amount of work involved in obtaining the root cuttings.

Hardening rooted cuttings

Hardening rooted cuttings refers to the development of plant resistance to environmental stress after rooting has occurred. For example, cuttings that have been rooted in a humid environment with moderate temperatures would be shocked if they were put directly in a dry, hot environment in full sun. There must be a transitional period to allow new roots and leaves to adjust gradually to environmental change.

The rooting period will vary from 2 to 16 weeks, depending upon plant species and the environment. The first step in hardening is to decrease the humidity by increasing the interval between mistings, and/or increasing the ventilation if in an enclosed rooting structure. After a gradual decrease in moisture, the light intensity can be increased gradually by moving the plants into areas receiving increasing amounts of direct sunlight. Plants that have been adequately hardened are more likely to survive when transplanted into larger containers or the landscape.

Layering

Layering is a relatively easy method of propagation by which new plants are formed while attached to the parent plant. The new plant receives nutrients and water from the parent plant until roots develop. This method of asexual propagation yields a large plant in a relatively short time, and is an excellent way to produce a small number of plants in the home landscape, or to propagate plants that are difficult to increase by other methods.

Layering outdoors is best performed during spring and summer months, although it can be done during any season of the year. Spring and summer layers are usually rooted and ready for transplanting in the fall or winter. Healthy, maturing branches that are growing vigorously and have been exposed to light should be chosen for layering since these usually have more food reserve (carbohydrates) and therefore root faster. Branches from pencil size to about ¾ inch (2 cm) in diameter are best for layering. It may be possible to select wood for layering that would normally be pruned when shaping the plant. The various types of layering are air, tip, trench, mound, and serpentine. Air and tip layering are the most popular methods.

Air layering

Air layering is commonly used for the propagation of fiddle-leaf figs, rubber plants, crotons, hibiscus, calliandra, oleanders, pandanus, camellias, azaleas, and magnolias. The first step in air layering is to remove leaves and twigs on the selected limb for 3 to 4 inches (8 to 10 cm) above and below the point where the air layer is to be made. The air layer is usually made at least 12 to 15 inches (30 to 38 cm) below the tip of the branch. The branch is wounded to induce rooting.

One method consists of removing a ½ to 1-inch (1 to 3 cm) ring of bark and, with a knife, scraping clean the wood underneath. This ensures complete removal of the cambium layer—a layer of cells between the bark and the wood. If the cambium layer is not removed completely, new bark may develop instead of roots. A second method of wounding involves making either a long slanting cut upward about one-fourth to one-half the way through the twig (Figure 2) or two small cuts on opposite sides of large branches or on branches having brittle wood. One cut should be slightly higher on the branch than the other and the cuts should not be too deep or the branch may break. The incision should be kept open by inserting a small chip of wood or toothpick to prevent the cut from healing over.

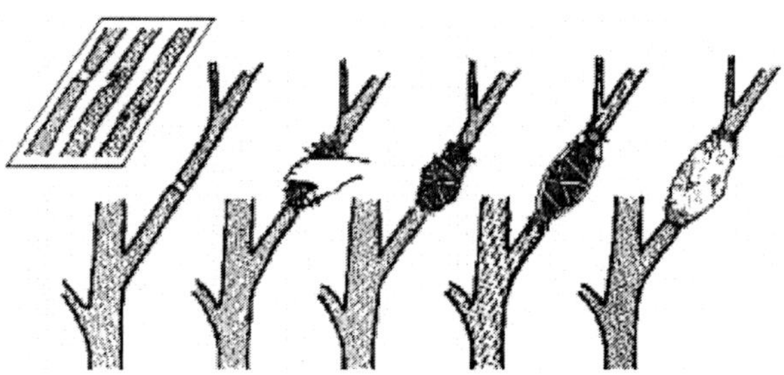

Figure 2. Air Layering, progressive steps in making an air layer.

A rooting hormone can be applied around and just above the wound on difficult-to-root plants to hasten rooting, but hormones are unnecessary for most air layering. The wounded area should be bound with a handful of moist sphagnum moss. Squeeze excess moisture from the moss before placing it completely around the stem at the wound. Tie the moss firmly in place with strong twine or fabric. Wrap the sphagnum ball with clear polyethylene film and tie securely with plastic covered wire or strong rubber bands above and below the ball to prevent the moss from drying. The ball should then be covered with aluminum foil or freezer paper to prevent excessive heat build up under the plastic.

When a mass of roots has developed in the sphagnum ball, the layered branch can be removed from the parent plant. When roots are visible through the plastic, the layer is ready for removal. It is best to allow the new plant to develop a larger root system in a container or protected holding area before planting it in open areas where high light intensities and dry conditions usually prevail. Layers removed during the growing season should be potted in containers and hardened much like the rooted cutting discussed previously. Layers harvested in winter can usually be transplanted directly into the landscape if adequate care is provided.

Tip layering

Tip layering is a proven means of propagating climbing roses, jasmine, abelia, oleander, and pyracantha (Figure 3).

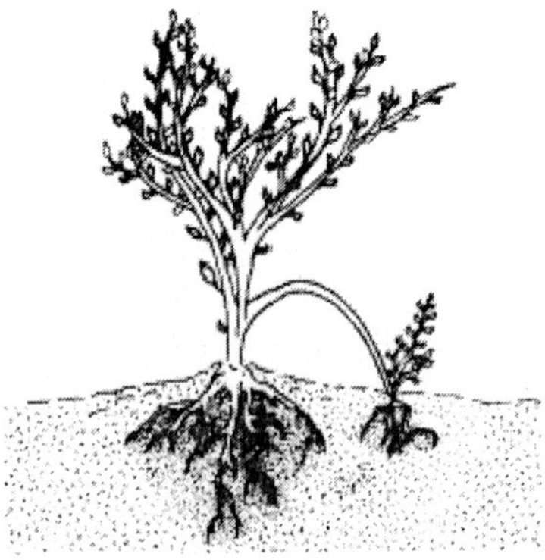

Figure 3. Tip Layering, an easy method of propagating plants around the home

Most plants with a trailing or viny growth habit can be propagated by this method. A low branch, or one that can be bent easily to the ground, is chosen. The bark is injured (in the manner previously described for air layering) about ½ to 1 inch (1.2 to 2.5 cm) along the stem and 4 to 5 inches (10.2 to 12.7 cm) back from the tip, and the injured area is anchored 2 to 3 inches (5 to 8 cm) in the soil. It is extremely important to keep the soil moist. Spring is the best time to tip layer, since the injured portion will develop roots during warm summer months. Spring layers can be cut from the parent and planted in late fall or left until the following spring. The layered portion should be checked for roots before removal from the parent plant.

Trench and serpentine layering

Trench and serpentine layering methods are similar to tip layering, except that a longer branch is placed in a trench and covered with soil. These methods produce several new plants from each layered branch (Figure 4).

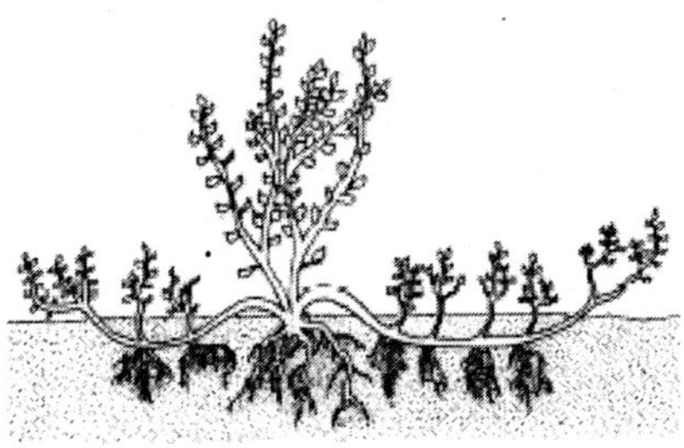

Figure 4. Trench Layering, this method is well adapted to the propagation of certain fruit and berry plants

Trench layering is useful on plants whose buds will break and start to grow under the soil surface. Willows, viburnum, and dogwood can be trench layered. The entire branch, except the tip, is placed in a trench and covered with soil. Serpentine layering involves burying every other bud, leaving the alternate bud above ground (Figure 5). This method requires plants with pliable or vining stems such as grapes, trumpet creeper, and confederate jasmine.

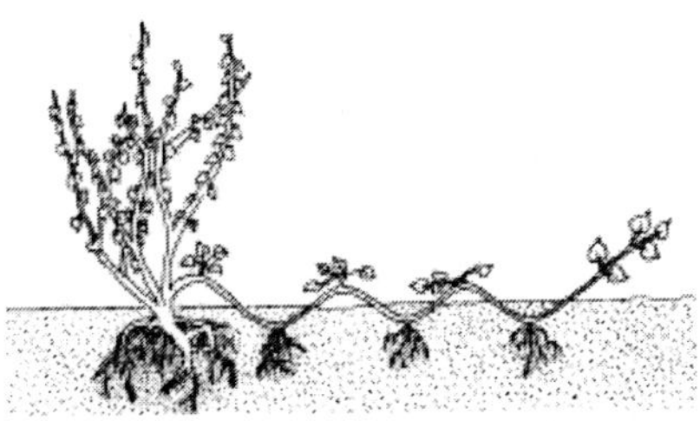

Figure 5. Serpentine Layering, alternate nodes along the branch are rooted, a method of securing a large number of plants from one branch

Mound layering

Mound layering can be used to propagate many of the heavy-stemmed or closely-branched plants such as Japanese magnolia, croton, flowering quince, calliandra, and tibouchina. Mound layering (Figure 6) is started in spring. The plant is cut back severely prior to spring growth; new shoots that emerge are wounded (as described for air layering) and soil is mounded around the base of the plant. Soil should be mounded up in several stages to a maximum of about 1 ½ feet (46 cm). Adding peat or sphagnum moss to the mounded soil helps when removing the rooted branches. It takes about one growing season to produce shoots that have rooted sufficiently for transplanting.

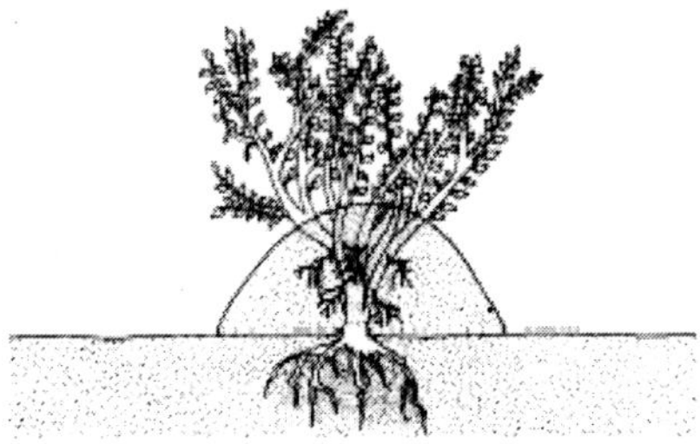

Figure 6. Mound Layering, showing roots forming in the soil mounded around the wounded stems

Division

Plants with a multi-stem or clumping growth habit, offshoots, or with underground storage structures such as rhizomes or tubers can be propagated by division. Division involves cutting large clumps into smaller sections, making sure that each smaller clump has an adequate amount of stems, leaves, roots and buds to survive transplanting. Ferns, orchids, daylilies, bulbous plants, nandina, and liriope are commonly propagated by division. Division is an excellent way to increase the area in the landscape covered with plants such as asparagus fern, confederate jasmine, and liriope. Each season dig the plants from a portion or all of the ground cover area, divide the clumps, and replant them into a larger area. Some plants can be pulled apart, but many must be cut. Transplant the separated clumps at the same depth they were growing originally Do not divide plants when they are flowering, but any other time during the growing season is suitable, as long as adequate care is provided after planting.

SEXUAL PROPAGATION

Seed propagation is the least expensive way to produce large numbers of new plants, but seedling characteristics are usually quite variable and this may be a disadvantage.

However, genetic variability offers an opportunity to select seedlings with new or different features. Seed propagation is a means of reproducing plants that are extremely difficult or impossible to propagate vegetatively. Most palms are grown from seed because they are single stemmed, and cuttings can not be taken without destroying the parent plant. Wax myrtle and native azaleas are usually propagated by seed because asexual methods are less successful.

Seed Collection and Storage

There are no firm rules for determining when seeds of selected plants are mature and ready for collection. Changes in size, shape, weight, and color of fruit serve as visual guides to seed maturation. For example, fruit of the southern magnolia are a brilliant red, and those of the pindo palm are orange and somewhat soft when mature. Fruit should be collected during this stage because after they fall to the ground viability may decrease. The period of seed viability for many subtropical and tropical plants is short—sometimes only 3 to 10 days. This situation is especially true for seeds coming from pulpy or fleshy fruit such as that of palms; these seeds should be planted immediately after they are harvested and cleaned.

Some seeds do not have to be planted immediately, but can be stored under controlled environmental conditions. Although optimum seed storage conditions differ with plant species, seeds should be separated from fleshy pulp as soon as possible after collection. Southern magnolia, ardisia, podocarpus, sea grape, and carissa are examples of plants with fleshy fruit. The flesh or pulp should be removed to aid drying, and because the pulp may contain chemicals that inhibit germination. Removal of the pulp by hand is possible for a small number of fruit, but alternative methods can be used for greater quantities of fruit. The pulp can be removed by allowing the fruit to soften in water and then scraping them over a wire screen. A blender with rubber tubing on the blades can also be used. Another method of pulp removal involves placing the fruit in a container with water and a small amount of coarse sand. Use a wire brush on an electric drill to stir the mixture and remove the pulp. Spread the clean seeds in thin layers in the sun or a warm room to dry.

Optimum storage temperature and seed moisture content vary with species, but generally seeds should be stored at 40°F (5°C) and in an environment with 30 to 35 percent relative humidity. Household refrigerators usually maintain temperatures suitable for seed storage, but the relative humidity may exceed that optimum for some seeds. Seeds can be stored in metal cans, plastic bags, or paper or aluminum foil lined envelopes. A protective fungicide treatment is advised for seed known to be susceptible to fungal diseases. Consult your local extension agent for recommended fungicides.

Seed Germination

Proper moisture, oxygen, temperature, and sometimes light must be provided for germination. Although optimum conditions differ with plant species, general recommendations can be made. Optimum temperatures for germination of most ornamental plant seeds are 75°F to 80°F (24°C to 27°C). A variation of 9°F (5°C) between day and night temperatures stimulates the germination of some species. The lower temperatures should be during the dark period.

The germination medium must hold adequate water yet drain freely. A mixture of equal volumes of peat moss and builder's sand is suitable, but other materials such as shredded sphagnum, vermiculite, and perlite used alone or in combinations are satisfactory. The particle size of a germination medium in relation to the seed size should be considered. A small seed positioned between large particles may dry rapidly even though the medium particles are moist, because there is inadequate surface contact between the seed and the germination medium.The medium should be sterile to prevent disease.

Damping-off, a common disease of seedlings, is caused primarily by the fungi *Pythium* and *Rhizoctonia*. Sterile propagation media can be purchased, or a small quantity can be sterilised in an oven (but the odor may be offensive). Heating a 2-inch (5 cm) layer of moist medium at 220°F (104°C) for 1 hour will kill pathogenic fungi. The medium should be moistened before the seeds are planted, and kept moist, but not too wet, for optimum germination. A fungicide treatment may be justified when specific seedlings are known to be susceptible to soil-borne fungi.

Seed should not be planted deeper than 1 to 2 times their diameter. Small seeds should be scattered over the germination medium surface or planted thickly in rows. Medium-sized seeds sown on the surface should be covered with a thin layer of shredded sphagnum or peat moss. Larger seeds should be planted at a depth less than their diameter since a 2- to 3-inch (5.0 to 7.5 cm) planting depth is maximum for any species. Coconut palm and cycad seeds are exceptions, and should be planted just under or level with the medium surface.

Seed Dormancy

Although seeds of many ornamental plants in Florida are ready to germinate as soon as the fruit matures, some seeds will not germinate until certain internal conditions are overcome. Such seed dormancy can be caused by an impermeable or hard seed coat. The seed coat may inhibit water movement into the seed or physically restrict embryo expansion. Seeds may also contain chemicals that inhibit water movement into the seed or physically restrict embryo expansion. Seeds may also contain chemicals that inhibit germination. Some chemical inhibitors are water soluble and can be leached from the seeds by soaking them in water. Other inhibitors must be degraded or modified by exposure to certain environmental conditions such as cold temperatures.

Seeds can exhibit dormancy due to an immature embryo, in which case proper storage allows further embryo development. Seeds can also be dormant due to a combination of these factors. Seed dormancy is nature's way of setting a time clock that allows seeds to initiate the germination process when conditions are suitable for germination and seedling growth.

Florida dogwood produces mature seeds in the fall, but conditions during late fall and winter are not suitable for seedling growth. Through evolution, the dogwood has developed a mechanism that keeps the seeds dormant until spring, when conditions are favorable for germination and seedling growth. Many Florida plants, especially tropical species, have no dormancy mechanism because conditions in nature at the time of seed maturation are usually conducive to germination and seedling growth. Plant propagators need only to provide a suitable environment for germination of these seeds as soon as they mature.

Dormancy caused by a hard seed coat can be overcome by breaking the seed coat. Scarification is the process of penetrating or cracking the seed coat barrier. Although acids and hot water treatments are sometimes used in commercial nurseries to break or soften the seed coat, mechanical scarification is most suited for the landscape gardener. Small numbers of seeds can be scarified by rolling them on a cement floor using a brick or board, by rubbing the seeds with sandpaper, or by cutting the seed coat with a knife. Mechanical devices may be purchased or constructed to scarify larger numbers of seeds. The seed coat should be dull in appearance after scarification, but not deeply pitted, or cracked enough to expose or injure the embryo. Scarified seeds will not store as well as nonscarified seeds and should be germinated as soon after treatment as possible.

Seeds of many temperate-zone plants require a cold period before they will germinate. This requirement is met by cold stratification - storing the seeds in a cold, moist environment. Seeds are mixed with moist sphagnum peat or vermiculite after a 12-to 24-hour soak in water at room temperature.

It is also advisable to spray the seeds with a protective fungicide treatment before putting them in refrigerated storage. The seeds should be stored for 2 to 6 months at 37°F to 40°F (3°C to 5°C). Temperatures in household refrigerators are usually adequate. Suitable containers for stratification are flats, trays, boxes, or cans that provide aeration, prevent drying, and allow drainage. Polyethylene bags no more than 0.004 inch (4 mil) thick may also be used. Seeds should be planted immediately after removal from refrigeration.

Seedling Establishment

Seed germination and early seedling development is best accomplished in a moist environment with moderate temperatures (75°F to 80°F or 24°C to 27°C). Although light is not required for germination of many seeds, high intensity light is necessary to produce stocky, strong seedlings. Low intensity light will result in weak and spindly, pale green seedlings.

Seedlings planted close together soon become crowded, resulting in slow growth and weak, spindly stems. Crowded seedlings must be transplanted with wider spacing into flats or individual or multi-celled containers. Seedlings can be grown in these containers until they are mature enough to transplant into larger containers or the landscape. Tender seedlings transplanted without a transition period into a hot, dry environment have poor survival rates. The environment in which seedlings are grown should be modified gradually until it is similar to the environment into which they will be transplanted. Watering frequency should be decreased gradually followed by a gradual increase in light intensity.

SMALL SCALE PROPAGATION UNITS

The key to successfully rooting cuttings and germinating seeds is a moist environment maintained at a favorable temperature. Environmental control is less important for other propagation methods, such as layering, because the mother plant provides some degree of support to the developing new plant. However, most cuttings and young seedlings are susceptible to environmental stress and will be successful only if an appropriate environment is provided.

An environment with a relative humidity near 100 percent will minimise water loss from cuttings and developing seedlings, although water loss is less critical for seedlings than cuttings. Cuttings cannot take water from the medium to replace that lost through the leaves, so if high rates of water loss occur, cuttings will dry-out. Temperature influences the physiological activity of plants. Excessively high or low temperatures injure plants or slow their growth and development, but temperatures in the range of 70°F to 80°F (21°C to 27°C) stimulate optimum growth and development for most plants.

The home gardener can provide a warm, humid environment for seed germination and rooting of cuttings by construction or purchasing small-scale propagation units. These units are inexpensive, require little attention, and are convenient to use in the home landscape or indoors. A propagation unit can be made from a terrarium or aquarium (Figure 7).

Figure 7. Aquarium used as Propagation Unit

These structures are usually constructed of glass or Plexiglass, but a suitable structure could be constructed of wood and glass or plastic. Approximately 2 to 4 inches (5 to 10 cm) of propagation medium can be placed in the bottom of the tank, and cuttings stuck or seed sown directly in the medium. Alternatively, 2 inches (5 cm) of gravel can be put in the bottom of the tank, and containers with propagation medium placed on the gravel.

Figure 8. Large Pot in Plastic Bag for Propagation

A glass or plastic cover should be put on the container after adequate moisture has been added. Large plastic pots and a plastic bag can be used to create a suitable propagation environment (Figure 8). Stick the cutting in a moist propagation medium in the container and add moisture as required. Place the whole container in a large, clear plastic bag. Wire hoops or stakes can be used to prevent the plastic bag from laying on the cuttings or seedlings.

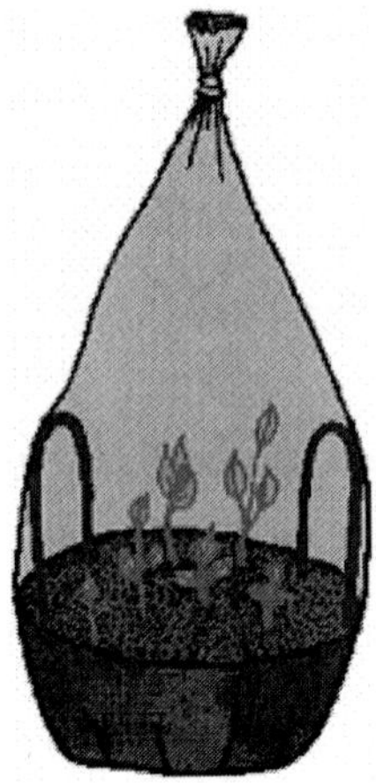

Figure 9. Plastic Bag Alone as Propagation Environment

A plastic bag alone can serve as a propagation environment (Figure 9). Simply place some moist propagation medium in the bottom of the bag, insert the cuttings, and tie the top of the bag closed. The placement of these simple structures is critical. Put the structures in diffused light and never in full sun. The temperature in these sealed units will rapidly become too high in full sun, and cutting or seedling injury or death will result.

Units kept indoors should be placed near a north window or under fluorescent lights for 12 to 16 hours per day. Temperatures of 65°F to 80°F (18°C to 27°C) should be maintained. Although these units are designed to prevent moisture loss, routine examination of the moisture level is suggested. Add moisture if no water has condensed on the inside of these units overnight, or if the propagation medium appears dry.

REFERENCES

Bir, Richard E., 1992. *Growing and Propagating Showy and Native Woody Plants.* University of North Carolina Press.

Bubel, Nancy. 1988. *The New Seed-Starters Handbook.* Rodale Press, Inc., Emmaus, Pennsylvania.

Clarke, Graham & Toogood, Alan, 1998. *The Complete Book of Plant Propagation.* Ward Lock Limited, Wellington House, 125 Strand, London.

Dirr, Michael A., 1987. *The Reference Manual of Woody Plant Propagation.* Varsity Press. PO Box 6301, Athens, Georgia 30604.

Garner, R.J., 1985. *The Grafter's Handbook.* The Royal Horticultural Society.

Hartmann, Hudson T., et.al. 1959. *Plant Propagation Principles and Practices.* Prentice-Hall, Inc., Simon & Schuster/A Viacom Company, Upper Saddle River, New Jersey.

Wells, James S., *Plant Propagation Practices.* 1985. American Nurseryman Publishing Company, 111 North Canal Street, Chicago,Illinois 60606.

6

Planting and Transplaning

A properly planted tree or shrub will be more tolerant of adverse conditions and require much less management than one planted incorrectly. Planting technique impacts water quality as it minimizes water, fertilizer and pesticide use. When making decisions on planting techniques, one should consider how the plant was grown in the nursery, the plant's drainage requirements, the soil type and drainage characteristics, and the availability of irrigation water. The plant should be specifically appropriate to the site, or the site should be amended to specifically fit the plant.

Horticulture researchers have estimated that 75% of the roots may be lost when digging field-grown nursery stock. Cultural practices by the nurseryman, such as root pruning, irrigation, fertilization, root-ball configuration, and digging techniques, influence the percentage of harvested roots. Water stress, due to removal of most of the water-absorbing roots, is the primary cause of transplant failure. Most water absorption capability within a transplanted root-ball results from very small diameter roots. These fragile roots are the first to suffer from excess water loss in newly transplanted landscape plants.

Landscape contractors and home gardeners can choose from a wide variety of plant material. Plants are grown by various production methods, e.g. bare-root, balled and burlapped, fabric container and plastic container. Some large landscape trees are mechanically dug with a tree-spade and placed in wire baskets. Each of these harvesting and growing techniques is acceptable, but requires a specific planting and management technique.

SITE SELECTION

Look at the space where you intend to plant the tree or shrub. Estimate the height and diameter of a tree or shrub that will fit there. Check to see if the soil stays wet, or if it drains quickly. Figure out how much space the roots will have. Ask yourself what role this plant will play in your yard. Will it be the center of attention? Is it part of a living wall that divides one part of the yard from another?

Once you know the function, soil type, sunlight, temperature, water, size, and root requirements for the plant, start thumbing through books and catalogues to find something to fit your need. Then, head out to a nursery to see what they have. Look for healthy, disease- and pest-free plants with well-formed root and branch systems. Resist any temptation to dig your own tree or shrub from the wild. Wild-dug plants often die after transplanting because they lose many roots.

If the site sometimes holds standing water, avoid plants that can't stand "wet feet" (Table 1). Once you know what kind of plant you're going to buy, you must decide if you intend to buy a bare-root, machine-balled, balled and burlapped, or containerised plant. How you treat the tree or shrub at planting time will partially depend upon how the roots were prepared for planting.

Bare-root. Advantages of planting bare-root plants are mostly economical. Plants are less expensive to produce because of the ease of harvesting, storing and shipping. Many species respond well to bare-root harvesting. A greater portion and longer roots are retained after harvesting and roots are easily inspected at planting time. Damaged roots can be trimmed and girdling roots can be removed before planting. Bare-root plants should be planted while they are completely dormant. Landscape-sized bare-root trees usually require staking.

Table 1. Commonly used landscape plants that do not tolerate wet soils.

Scientific Name	*Common Name*
Abies concolor	White Fir
Acer saccharum	Sugar Maple
Circidiphyllum japonicum	Katsura tree
Cladrastis lutea	Yellowwood
Cornus florida	Flowering Dogwood
Fagus spp.	Beeches
Hedera helix	English Ivy
Pinus strobus	White Pine
Quercus rubra	Red Oak
Rhododendron sp.	Rhododendrons & Azaleas
Taxus sp.	Yews
Tilia cordata	Littleleaf Linden
Tsuga canadensis	Canada Hemlock
Vinca minor	Myrtle

Machine-balled Plants. Nursery workers dig these plants bare-root, then pack the roots in sphagnum peat moss or wood shavings and wrap them in plastic. Handle them carefully to avoid breaking roots. These, too, should be planted while they're still dormant.

Balled and burlapped Plants (B & B). Larger landscape plants are traditionally harvested as "balled and burlapped" (B&B). A major advantage of B&B plants is that soil types can be matched, thereby reducing any interface problems that might inhibit water movement between the rootball and surrounding soil of the landscape site. There is an acceptable, standardized formula for sizing rootballs, which is the American Standard for Nursery Stock. The main disadvantage of B&B material is that a large portion of the roots may be severed at harvest time. The amount of roots harvested depends upon soil type, irrigation practices and root pruning during the production period. Plants moved B&B are subject to seasonal constraints. The most favorable seasons are when transpiration demand is low and root generation potential is high, such as in fall, winter and early spring. With the much-reduced root system, water is a critical element in the successful transplanting of B&B material.

Container-Grown Plants: The advantage of using plants grown in containers is that 100% of the roots are in the container. Thus, the plant goes through limited transplant shock if given adequate follow-up care. Container-grown plants can be planted into the landscape year-round. Plants produced in containers, in a soilless medium (usually bark and sand), are much lighter than B&B material. This is very helpful to home gardeners who may not have large equipment to handle the heavy plants.

The main disadvantage of container-grown plants is the possibility of deformed roots. "Rootbound" plants have roots circling inside the container. The entangled roots are a physical barrier to future root growth and development. If this condition is not corrected at planting time, the plant may experience slow growth and establishment because of the girdled roots. Some form of root mass disturbance is recommended before planting.

A relatively new production system is the use of fabric containers or bags. Plants are grown in the bags, placed in the ground, with a soil backfill. The advantage to this production technique is purported to be a means of harvesting a greater number of roots while using field production practices. The fabric must be removed at transplanting time. This can be somewhat of a problem when the roots have become attached to the walls of the bag, or if roots have escaped through the fabric.

WHEN TO PLANT

You should plant most trees and shrubs early in the spring, just before or as new growth starts. Certain tree species essentially demand spring planting, because they establish new roots very slowly (Table 2).

Table 2. Some slow-to-root trees that should only be planted in spring.

Scientific Name	Common Name
Acer rubrum	Red Maple
Betula spp.	Birches
Chamaecyparis nootkatensis	Nootka False Cypress
Cornus florida	Flowering Dogwood
Crataegus spp.	Hawthorns
Koelreuteria paniculata	Goldenraintree
Liriodendron tulipifera	Tulip Tree, Tulip-poplar
Magnolia spp.	Magnolias
Nyssa sylvatica	Black Gum
Populus spp.	Poplars
Prunus spp.	Stone fruit (Peach, Cherry, etc.)
Pyrus calleryana	Callery Pear,
Including	'Bradford'
Quercus alba	White Oak
Quercus coccinea	Scarlet Oak
Quercus macrocarpa	Bur Oak
Quercus phellos	Willow Oak
Quercus robur	English Oak
Quercus rubra	Red Oak
Salix spp.	Willows
Tilia tomentosa	Silver Linden
Zelkova serrata	Japanese Zelkova

Most trees and shrubs can also be planted in fall. If you've bought plants in containers or balled and burlapped, you're even less time-bound. Because the soil stays with the roots, you can plant them any time the soil can be worked. Try to buy plants just before you intend to put them in the ground. Until they're planted, you'll need to protect the roots from damage, such as drying and overheating.

You're preparing a spot where a tree or shrub may stand for decades, so be sure it is well-drained and well-aerated. Roots need water and air to grow properly. Soil types vary not only between regions, but also between two spots in a yard. Know your soil conditions before you plant. Check subsurface drainage by digging a hole and filling it with water. If the water doesn't drain away within two hours, you either need to improve the subsurface drainage or select only those plants that can tolerate wet soil. Improving soil drainage and aeration on a large scale is difficult and expensive. You may need to fill and regrade water-collecting areas, install drain tiles, or incorporate organic matter in conjunction with deep spading or plowing. If you can't extensively rework a poorly drained site, be sure to select plant species that can tolerate soggy or clayey conditions.

HOW TO PLANT

Smaller plants live better and establish faster than large plants and are more economical. Many consumers, on the other hand, want the "instant" landscape look. Demand for large, landscape-size trees has certainly increased over the last decade. With large mechanical digging equipment, 6- to 8-inch diameter trees can be moved. Large diameter trees are often trans-planted for instantaneous effect, but post-transplant stress and costs increase with the size of the tree.

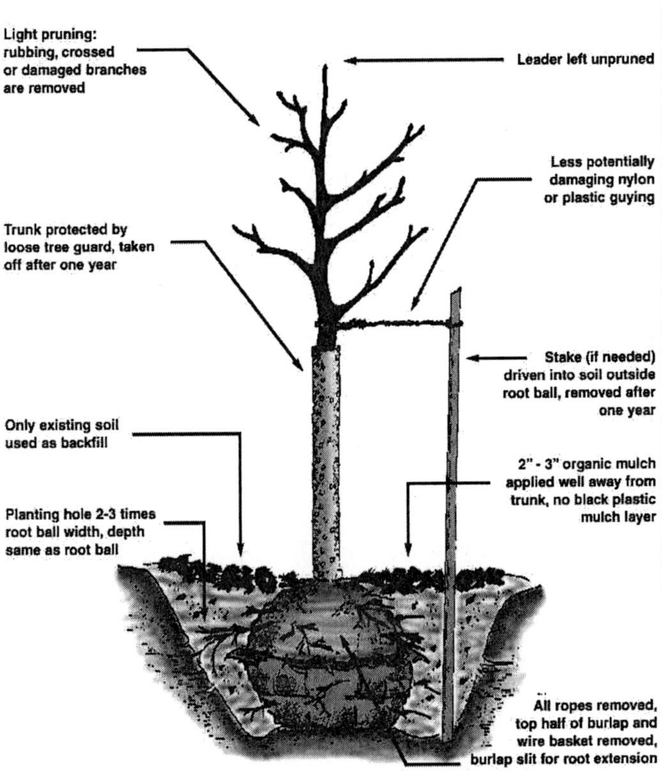

Figure 1. A propoerly planted tree

1. Plant understory trees, like dogwoods, in shade, not in full sun.
2. Do not plant water-loving trees, like willows, near septic fields or water sources. Relocate or change species.
3. Align trees for wind screens where they will block winter wind, not where wind will funnel around or between them.
4. Avoid planting trees with leaf or fruit "litter" near pools, decks and paving. Locate where they shade and self mulch.
5. Avoid planting evergreens where they block winter sun. Instead, plant deciduous trees for summer shade and winter solar radiation.
6. Avoid planting trees too close to the house. Plant further away, or plant smaller trees and shrubs.
7. Cars often scrape and bruise tree trunks. Locate trees away from parking spaces or areas where they can be vandalized.
8. Branches can create hazards or block views. Locate trees away from street corners, or vehicle and pedestrian traffic.
9. Avoid planting trees directly under utility lines. Locate trees away from lines or plant smaller, low-growing trees or large shrubs.
10. Plant trees away from sidewalks to allow root expansion without pavement buckling.

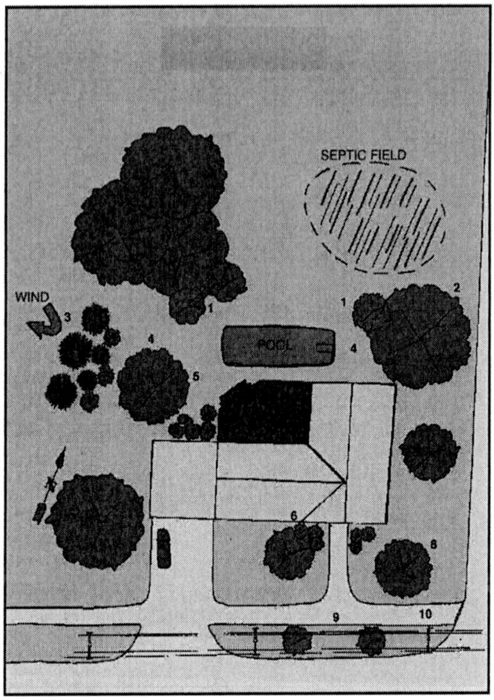

Figure 2. Selecting correct location

Planting Procedures

Correct planting technique begins with the loading of the plant at the nursery or garden center. Home gardeners and landscapers should be very careful with plant material. Always protect the roots, stems and foliage during transport. The plant tops should be shielded from winds. Never pick up a plant by the trunk. Trees are particularly vulnerable to damage if growth has started. In the spring the bark is easily injured. B&B trees are very susceptible to this type injury because of the weight of the root-ball. Lift plants from underneath the rootball with the appropriate equipment. Container- grown plants should be handled by the container and never by the tops of the plant. If plants must be held or stored on the landscape site, it is best to place them in a location protected from the wind and sun. Do not let the roots freeze or dry out during this time. If the delay in planting is more then a few days, one should "heel in" B&B material by covering the roots with bark or some other mulch. Supplemental irrigation is critical for the nursery stock during the growing season.

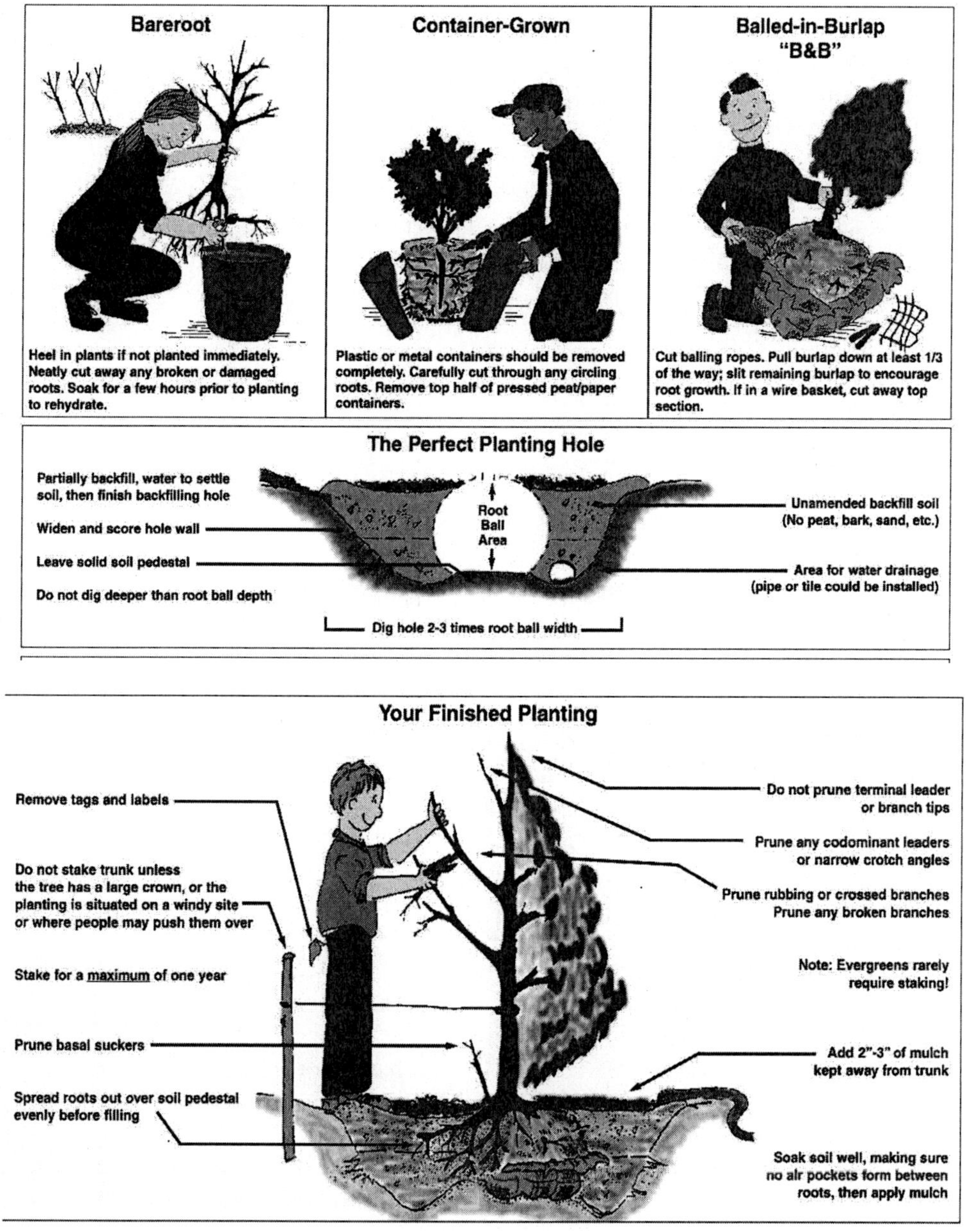

Figure 3. Planting trees and shrubs: Types of nursery stock

Preparing the hole

A current trend in landscape design is to plant trees and shrubs in large beds. When this design concept is followed, preparation of the entire plant bed area and not just individual holes is recommended. In many urban areas, gardeners will find that the soils are compacted and sometimes poorly drained. In these situations one should create a good rootzone by amending the beds with a sandy-loam topsoil and aerifying the soil as deep as possible. The addition of organic matter provides little or no advantage to the planting hole in good soils. Backfill should, in most cases, be the soil removed from the planting hole: "what comes out...goes back in". This is especially important for B&B material and bare-root planting stock. An exception to this would be where entire beds can be amended to create an homogeneous root-zone. The organic matter, e.g. compost or composted pine bark, is uniformly mixed with the soil. This makes room for future growth and increases aeration to the backfill.

In very poorly drained soils, drain tile under the beds is necessary. If a french drain or tile drain is installed, be sure that it drains downhill at a 2% minimum slope and there is an outlet on the downhill side. When setting plants, be certain to plant them high. If the poor drainage condition cannot be corrected, don't plant a tree or shrub in the area, unless it can tolerate these conditions.

In loose, well-drained soil, dig the hole two times the width of the root spread, soil ball, or container size of the plant. Dig as deep as the root system, but not much deeper, so that when you have finished planting the tree or shrub is just as deep in the new location as in the old. Flatten the bottom of the hole and stand in it to firm the soil so that the plant doesn't settle in deeper later. If your plant is bare-root, you can make the shape of the bottom of the hole fit the shape of the root system. In most cases, use the same soil that came out of the hole to backfill; mix topsoil and subsoil together. To avoid burning roots, do not add dry fertilisers or fresh manure to the backfill mix.

If you are planting shrubs or small trees in very well-drained or light, droughty soils, thoroughly mix in one part of a good grade sphagnum peat moss with two parts soil. Such soil amending only marginally helps large trees with extensive root systems. Do not amend backfill soil on a heavy soil site. If you plan to stake the tree, pound stakes into the ground now, before you plant, to avoid damaging roots. Once the hole is dug, planting instructions will vary according to the type of root preparation your tree or shrub got from the nursery.

Planting Bare-Root and Machine-Balled Stock

On bare-root stock, carefully remove the moist packing material and examine the roots. Cut off damaged roots with a sharp knife or pruners. Soak plant roots in water for several minutes immediately before planting, but don't let roots stand in water for more than an hour. You must always protect the plants roots from drying. Even while you're digging the hole, keep roots covered with damp burlap, moist sphagnum moss, or other material to avoid exposing roots to sun and air. For machine-balled stock, dig the hole before you remove the plastic wrap. After you take the wrapping material off of the root ball, pull the peat ball apart gently to let roots contact soil.

For both bare-root and machine-balled plants, be sure the plant sits at exactly the same level in the new hole as it did where it grew before. Stand it in the center of the hole and carefully backfill with soil. Work the soil in and around the roots, then firm it with your hand. Continue filling the hole until it is three-fourths full. Gently tamp the soil with your feet, but don't pack the soil or break roots.After the hole is three-fourths full, fill it full of water and let the water drain. This settles the soil and eliminates air pockets around the roots. Do not pack the soil after it is watered. Straighten the plant if it's crooked, and finish filling the hole. To catch and hold rain water and to make watering easier, form a 2- to 3-inch rim of soil in a circle 2-3 feet larger than the diameter of the hole (Figure 4).

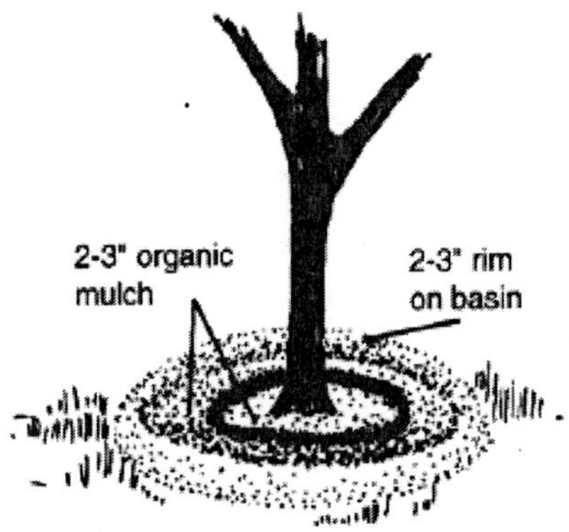

Figure 4. Finish your Planting Job by Creating a Watering Basin and Mulching

Prepare a fertiliser solution with a water soluble fertiliser (20-20-20 or similar analysis). Use the rate recommended on the label. Fill the basin with the fertiliser solution, usually 2 to 3 gallons for each plant. Additional fertiliser is not needed during the first growing season.

After watering, add 2- to 3-inches of bark mulch or other coarse material over the entire watering basin. Keep all mulch away from the trunk of the tree and the stems of shrubs. For the first year, maintain the soil rim around the basin to catch rain or irrigation water.

Planting Balled and Burlapped Stock

Always handle balled and burlapped stock by the soil ball, never by the trunk or crown of the plant. Stand the plant in the prepared hole so that the top of the soil ball is level with the surrounding soil surface, never deeper. Backfill the hole three-fourths full. Cut all twine or wire away from the top of the soil ball and the trunk. Completely remove wire baskets, if practical. However, removing them from large soil balls may cause the ball to fall apart. Evidence suggests that wire baskets do not cause long-term problems for plant growth, as long as the wire is well below the ground. Roll back the burlap to below the soil surface. Fill the hole with water to settle the backfill. Finish by filling the hole, but do not pack or tamp this soil. Finally, prepare a basin, then mulch and fertilise as described for a bare-root plant.

Planting Containerised Stock

Dig the hole before you remove the container. Also, thoroughly water the containerised plant. You must remove metal or plastic containers completely. Plants have likely been in those containers for a full growing season and have a dense root ball. Turn the container upside-down and give the rim a sharp tap.

The root ball should fall out in one piece. Before you put the root ball in the prepared hole, cut any long roots that completely encircle the root ball. Gently pull other roots away from the ball and spread them out. This will allow you to place backfill soil directly around those roots. Finish planting, mulching, and fertilising as described for B & B stock.

The root ball of plants in papier-mache or other degradable containers will be loose because the plant has been in the pot for less than a full growing season. The soil will probably fall away from the roots if you take them out of the pot. Instead, position the pot in the prepared hole the correct depth. Tear away any part of the pot that extends above the soil line. Use a sharp utility knife to slash the pot vertically in five or six places. Backfill immediately while the form of the pot and soil ball are undisturbed.

Planting in Heavy Soil

If you want plants to survive in heavy, clayey soil, you must pick plants that can tolerate these conditions. You won't have as many plant species to choose from, but if you choose well-adapted plants they should survive with minimal maintenance. Dig the hole at least three times the diameter of the root ball. Use the soil from the hole to backfill. You may be tempted to dig a big hole for the tree or shrub, plant, then fill in with lighter soil or an amended soil mix. However, if you do, you likely will create a bowl that catches and holds too much water, suffocates roots, and kills the plant.

If you absolutely must plant species of small ornamental trees or shrubs that require excellent drainage in poorly drained soils, create a raised bed for them. Plan for a bed at least 6- to 8-inches high and at least 4 feet wide for a shrub or 8 to 10 feet wide for a small ornamental tree. To build a raised bed, rototill or hand spade the existing soil. Then place a 3- to 4-inch layer of well-drained soil on top. Spade or rototill the added soil in place. Follow with a second 3- to 4-inch layer of soil and a final pass with the rototiller. You can build a wall around the raised bed to prevent the soil from eroding away; however, where space permits, slope the beds outer edges to the original level of the soil.

Planting in Excessively Sandy or Light Soil

For extremely sandy "light" soil, prepare a backfill mixture of one part sphagnum peat moss and two parts original soil. You can dig the hole larger than generally recommended, but, as always, set the plant no deeper than it had originally been growing. Backfill with the prepared mix and add at least a 3-inch layer of mulch outward from the trunk to a point 6 inches beyond the width of the planting hole. Water thoroughly once a week. Because sandy soil does not retain nutrients well, be sure to include water-soluble fertiliser in the first watering and again once or twice during the first season.

PROCEDURES AFTER PLANTING

Pruning

Planting time is excellent for pruning off diseased or damaged branches, basal shoots, and limbs with extremely narrow crotch angles. For shade trees, you can take this time to select major scaffold limbs to keep, then prune out excess branches.

Staking and Guying

To keep trees from tipping and the roots from moving too much, stake any bare-root tree larger than 6-feet tall, and balled and burlapped trees larger than 10-feet tall. Small trees usually don't need the support. Drive stakes into the undisturbed ground before you backfill the hole, to be sure you don't drive the stakes through the root ball and damage the roots. Larger trees should be guyed. To guy a tree, use three wires attached to three stakes. Thread a one foot piece of hose on each wire and have the hose-covered bit of wire loop around the tree at the lowest branch crotches.

Make loops around the trunk very loose to avoid damaging the tree. Drive stakes into firm soil at least 18 inches outside the perimeter of the planting hole. Anchor the loose ends of the wires securely to stakes. Various fabric straps and rigid staking systems are available that may be easier to use than hose and wire. Don't forget to remove the stakes and guys before the wire girdles the trunk. Generally, remove supports after one growing season for a 1-inch diameter tree, two seasons for a 2-inch diameter tree. For larger trees, guys may need to remain for three seasons or more. They should be inspected annually and adjusted to prevent trunk girdling.

Trunk Wrapping

In the late fall, wrap newly planted trees, especially thin barked trees like red maple, with a light-colored, commercially available tree wrap to provide winter protection from sunscald. Remove the wrapping material in spring. Wrap trees each fall until the bark is rough and corky. Start the wrap at the base of the tree, and extend it to the first limb. Spiral the wrap around the trunk with each turn overlapping the previous turn by half the width of the material. Secure the wrap with tape, twine, or by looping it back on itself. You may want to surround the lower part of the trunk with wire or plastic guards to discourage rabbits and rodents that eat bark.

Watering

If you want your newly planted trees and shrubs to survive their first year, you've got to be sure they get the right amount of water. Overwatering is just as harmful as underwatering. How often and how much you water depends upon your soil type and the amount of rainfall. On well-drained soils, apply 1 inch of water per week in summer and fall. On sandy soils, give plants at least 2 inches of water per week, preferably in two 1-inch applications. Plants in poorly drained, clay soils need less frequent watering.

Water regularly and supplement brief rain showers. If you get a heavy rain, you may still need to water the following week. Often much of a heavy rain runs off. If you use a lawn sprinkler, put a straight-sided can near the tree or shrub and water until the can contains one inch of water.

TRANSPLANTING

Transplanting or replanting is the technique of moving a plant from one location to another. Most often this takes the form of starting a plant from seed in optimal conditions, such as in a greenhouse or protected nursery bed, then replanting it in another, usually outdoor, growing location. Botanical transplants are used infrequently and carefully because they carry with them a significant risk of killing the plant.

Transplanting has a variety of applications, including:

– extending the growing season by starting plants indoors, before outdoor conditions are favorable;

- protecting young plants from diseases and pests until they are sufficiently established;
- avoiding germination problems by setting out seedlings instead of direct seeding.

Different species and varieties react differently to transplanting; for some, it is not recommended. In all cases, avoiding transplant shock—the stress or damage received in the process—is the principal concern. Plants raised in protected conditions usually need a period of acclimatization, known as hardening off. Also, root disturbance should be minimized. The stage of growth at which transplanting takes place, the weather conditions during transplanting, and treatment immediately after transplanting are other important factors.

Sometimes you must move a small tree or shrub from one spot to another in a yard. It's best to think ahead. Two years before the move, start pruning roots in early fall while trees are still growing. This gives the plant the best chance for survival following the move. If you don't root-prune the plant before the move, it may still survive, but the chances of success are reduced. To prune roots of small trees, first measure the trunk diameter at your waist height. Then, measure out 9 inches from the trunk for each inch of diameter.

A tree with a diameter of 1-1/2 inches should be root pruned 13-1/2 inches from the trunk. For shrubs, make the root-pruning cut half the distance of the radius of the branch spread of the shrub. For example, a shrub with a spread of 36 inches from the center to the outermost branches should be root pruned to 18 inches. When plants are dormant during the first year, cut straight down around two quarters of the plant on two opposite sides. Use a sharp spade to cut 18 inches deep. During the second year, cut around the remaining two quarters of the perimeter of the dormant plant. This two-year process cuts off many long roots and encourages new roots to grow in the soil ball. Move the plant during the third season.

When you're ready to move a plant, tie the branches to the central trunk. Then start digging 6 inches further away from the trunk than the root-pruning line. Dig a hole 18 inches deep. When the circular hole is completely finished, gently rock the plant from side to side. Cut the roots on the bottom of the soil ball and slide burlap under the ball. Grasp all four corners of the burlap, and lift the soil ball out of the hole. Wrap the burlap tightly around the soil ball. Tie the burlap so it cannot get loose in handling. Handle the plant carefully, by lifting the soil ball, not the trunk.

Before transplanting a woody plant, evaluate whether or not the tree or shrub is likely to be a successful transplant. Transplanting stresses trees and shrubs. Such stress may cause plants to die or to become unattractive. Plants which are already in advanced stages of decline are especially likely to succumb to transplantation stress. Often a young nursery-grown plant will resume growth sooner than an older transplanted tree or shrub and will provide more long-term benefits in the new planting location. Shrubs have better transplant tolerance than trees, deciduous plants better than evergreens, shallow rooted species better than deep rooted species, and younger plants better than older plants.

Some species may survive transplanting any time during the year when the ground is not frozen, but woody plants are preferably moved in the spring after the ground thaws and before the buds on the tree or shrub begin to swell. They may also be moved in the fall after leaf drop but before the ground freezes. Fall planting should take place soon after leaf drop, providing time for new water absorbing roots to develop before the soil freezes. This is often difficult to estimate in the Northern Plains.

Since evergreens are especially prone to winter browning if planting is delayed until shortly before the ground freezes in the fall, they should be moved late in the summer to early fall. Properly applied antitranspirants may help reduce the effects of winter desiccation in some species. Fall transplant success may be increased by transplanting hardy plants into sites with good soil moisture and wind protection. Woody plants that are transplanted in late spring and early summer, when shoot growth is at its peak, tend to show the greatest transplant injury.

There are great differences in the environmental requirements for each tree and shrub species. Only transplant a tree or shrub where light, moisture, soil pH, and wind exposure are appropriate for the particular species. All plants require space for root and crown development; therefore, consider mature plant size when planting trees and shrubs.

Soil characteristics are often limiting factors for woody plant survival in a given area. Sometimes the soil is inappropriate for tree growth and will require improved drainage or amendments before trees and/or shrubs are planted at the given location. A soil test should be completed in areas where soil quality is questionable. Never allow plant roots to become dry during the transplanting process. Water all woody plants two to three days before digging if the soil is dry. Prior to digging, shrubs and trees with low branches should have these branches tied up to prevent injury during the digging, transporting and planting operations. Marking one side of the trunk will allow a tree to be placed in the same orientation at which it grew in its original location. Consistent orientation may help to prevent sunscald injury to stems. A sharp spade should be used when digging trees to assure root wounds are clean cut. Although leaving a soil ball attached to the root system will cause less root injury, soil is heavy and sometimes it is more convenient or even necessary to transplant a tree without a soil ball.

Deciduous trees with a stem diameter of less than 1 inch and small deciduous shrubs may be dug either bare root or with a soil ball. Larger plants should only be dug with soil attached. Bare root transplanting should only be done in the spring and care must be taken to prevent damage to roots when removing the soil. Most shrub species require a root ball diameter of about two-thirds of the branch spread. The soil ball for trees should be a minimum of 12 inches for each 1 inch of trunk diameter.

Large shrubs and trees should have a trench dug deep enough to get below all of the major roots (usually 15 to 24 inches). The trench should be dug completely around the tree or shrub to be transplanted. This will provide the angle necessary for the spade to undercut roots directly under the soil ball. Shrubs under 4 feet tall do not typically require trenching because the soil balls are small enough for the spade to make the undercut without a trench. All roots around the plant must be severed before any lifting takes place. If the plant is removed from clay soils, any glazing of the soil ball should be roughened before burlapping or potting.

Trees and shrubs that have been dug for transplanting should be planted as soon as possible. Cover a root ball with damp material which will retain moisture (burlap, peat moss, canvas, plastic, etc.) until planting. Plastic should only be used in shaded areas for less than a day or heat injury and/or root suffocation may occur. When a tree or shrub is stored, it should be protected from direct sunlight, winds, and temperature extremes. If any woody plants cannot be planted for more than a week, their roots should be covered with a mulch or moist soil and the plants should be placed in a shaded area. In all cases root systems should not be allowed to dry out. Dry roots can severely decrease the potential for transplant success. Trees and shrubs must be protected when transporting to a planting site. Covered trucks and vans are best, but if a pickup truck is used, a tarp must be in place to protect the plant canopies and roots from drying winds in transit.

Transplanting Large Trees

Special considerations are necessary when moving large trees. If trees are over 3 inches in diameter, special equipment is often required to transport the tree. Depending on the size of the tree and the technique used, the equipment may include hand carts, winches, tree spades, or cranes. If trees will be transported on a truck, precautions must be taken to ensure that they will clear power lines, bridges, and other obstacles. Permits may be required to transport large trees on some public roads. For trees not grown with the benefits of nursery production, root pruning the trees for two or more years prior to transplanting may prove beneficial in reestablishment.

When hand digging, the techniques are the same as for smaller trees. Hand dug large trees may be balled and burlapped (B&B) or boxed. Larger B&B trees should have additional support provided by rope or wire. Chicken wire is a convenient material that can be wrapped outside of the burlap to support root balls. If a crane is used to pull the B&B or boxed tree from the hole, lift from the bottom of the root ball. Ensure that the trunk is heavily padded if a cable must be secured around it to balance the tree during removal. Since there is potential for severe bark injury, cables should be secured around the trunk only when they are absolutely necessary to stabilize the tree for lifting and transporting.

Boxing trees is sometimes preferable to B&B. Boxes will hold the root ball more securely than burlap. This is helpful in sandy soil or when trees are held for extended periods of time. Trees are dug in the same way as B&B, only the root ball is formed to fit snugly into a box. After the lateral roots are severed, the sides of the box are secured in place. Then the descending roots are severed and the bottom of the box is secured before lifting from the hole. Large boxes require heavy metal bands or other support to hold them together. Boxes may also be used to transplant trees which are larger than mechanical spades can successfully transplant. These trees should be side-boxed with the root ball diligently monitored to prevent drying out for at least three months prior to severing the descending roots and securing the bottom of the box.

Tree spades have become increasingly popular and are commonly used by professionals to move trees quickly and inexpensively. Only individuals properly trained in the maintenance and operation of tree spades should use them. Sharp blades reduce damage to roots during transplanting. Crushed or shredded roots caused by dull blades will develop more dieback than clean cut roots. Large trees should not be transplanted with root balls smaller than 12 inches in diameter for each 1 inch in trunk caliper. If multiple trees are being transplanted, all of the trees may be dug and stored B&B or boxed before transporting them to the new site. Increased transplant success may be achieved by tilling an 18 to 24 inch wide band adjacent to the outer edge of the root ball. This allows easier penetration of roots from the transplant ball into the adjacent soil area. If planting into clay soils, the sides of the hole should be roughened with a rake or shovel. When tree stability is questionable, guy at three locations, using non-abrasive materials, only until the tree has adequately reestablished anchorage through new root development. Before moving a large tree, keep in mind that smaller trees of a particular species typically transplant better and catch up in growth to larger trees of the same species. A general rule is for each inch in caliper, a year is required for transplant recovery; therefore, a 4 inch caliper tree may require four years to recover from the transplant procedure before normal, active growth resumes.

REFERENCES

Ian McKiel, Erin McManus and Mary McDonald, students, University of Minnesota Department of Horticultural Sciences (2006). "Moving and Transplanting Trees with a Tree Spade". http://www.sustland.umn.edu/implement/treespade.htm. Retrieved on 2006-12-28.

University of Minnesota Extension Service 1999. *Planting and Transplanting Trees and Shrubs*. http://www.extension.umn.edu/distribution/horticulture/DG3825.html. Retrieved on 2006-12-28.

7

Pruning of Plants

Pruning is necessary if the value and quality of the trees and shrubs in the landscape are to be maintained. The main reasons for pruning are removal of dead, diseased or injured branches; to improve the form, shape or size; to rejuvenate older plants and for safety and convenience. Generally, the best time to prune woody plants is in the early spring while they are still dormant. Exceptions to this rule include plants that will "bleed" if pruned in winter or early spring. Although this loss of sap causes no harm to the plant, it is a source of concern for many homeowners. To prevent sap loss in the spring, these plants should be pruned when they are fully leafed out. These would include maples (including boxelder), birch, black walnut, elms and honeylocust.

Trees that bloom on old wood should be pruned immediately after blooming. The flowering buds of these plants are produced the previous growing season. These plants include: forsythia, lilac, viburnum, mockorange and spirea.

Evergreens can be pruned in the early spring (March-May) or in early summer after any new growth has hardened. Pines probably have the most particular pruning requirements. In the early spring they produce new growth known as "candles." These candles can be pruned when they have elongated but before the needles fully emerge. Junipers, arborvitaes and yews can be pruned anytime from mid-April to mid-August. Pruning later would likely leave unhealed wounds that could result in winter damage.

Pruning is necessary to maintain a healthy, vigorous tree or shrub. Specifically, pruning is practiced to:

- *Maintain or reduce plant size.* Pruning can prevent a plant from overgrowing its space in the landscape and eliminates the need for drastic cutting of crowded, overgrown plants. It can allow for growth of plants under or adjacent to the pruned plant. It can also serve to reduce leaf area on newly planted trees and shrubs. This promotes survival through transplanting and consequent root loss.

- *Remove undesirable growth.* Pruning can encourage plant vigor through the removal of weak, overcrowded growth. Such thinning often improves the visual balance or symmetry of the plant.
- *Remove dead, diseased, or broken branches.* Pruning will aid in maintaining the shape, vigor, and health of the plant.
- *Stimulate flowering and fruiting.* Removal of the current year's old, faded flowers and fruit clusters will promote flower buds for the following season.
- *Rejuvenate and restore old plants to vigorous growth.* Proper pruning can restore a youthful, natural growth habit in certain overgrown shrubs.
- *Prevent damage to life and property.* Pruning can minimise the hazard of limbs interfering with power lines or overgrowing structures. It can also remove weak crotches before limbs break in strong winds and open blocked sight lines caused by overhanging limbs at driveways or street corners.
- *Shape plants in an artificial form.* Pruning and shearing can be used to shape plants as hedges or for rigidly formal espaliers or topiaries.

MYTHS ABOUT PRUNING

There are a number of myths and misconceptions about pruning which should be laid to rest.

- *Pruning is difficult.* Pruning is straightforward if one knows a little about how the plant grows and what it should look like when the process is complete.
- *Plants will die if pruned at the wrong time of year.* Plants may be injured, but seldom, if ever, are they killed by poorly timed pruning.
- *All pruning must be done during the winter.* Actually, many plants are best pruned during the growing season.
- *Topping shade trees will keep the trees from causing damage to the home.* Shoots which grow after topping are weaker than the original limbs. They will be more likely to split off and cause damage unless they are removed every few years. Also, wood rots are more likely to be a problem in topped trees, resulting in poorer tree health and greater likelihood of limb breakage.
- *Removing a tree is a crime against nature.* If a plant is in the wrong place, from a functional or aesthetic viewpoint, it is by definition a weed and can be removed. This is especially true when a tree must be mutilated beyond recognition to eliminate the problem it is causing.

- *Most trees need pruning.* Actually, mature trees seldom do. Young trees usually benefit because pruning helps in establishing the basic branch structure and in overcoming transplanting shock.
- *Hedge shears are all you need to prune shrubs.* Hedge shears are intended to prune hedges, only! Using them on shrubs not intended as hedge plants destroys the natural grace and beauty of the plants.
- *Anyone with a pickup truck and a chain saw is a qualified pruner.* Indiana has no licensure for tree pruners, thus some individuals doing pruning work may not be knowledgeable or skilled in proper techniques. Never hire someone who stops and tells you that your plants need pruning and that he will do it right away.
- *All cut surfaces must be treated with tree paint.* While long recommended, the evidence is conflicting on the use of tree paint. Largely, its use should be thought of as cosmetic, helping to hide light-colored scars.

In nature, certain meteorological conditions such as wind, snow or seawater mist can conduct a natural pruning process. The purpose of anthropomorphic pruning is to shape the plant by controlling or directing plant growth, to maintain the health of the plant, or to increase the yield or quality of flowers and fruits. In general the smaller the wound (smaller the branch that is cut) the less harm to the tree. It is therefore typically better to formative prune the tree when juvenile than try to cut off large branches on a mature tree. There are also differences pertaining to pruning, involving roses, shrubs, hedges, fruit tree or amenity trees. If a shrub is incorrectly pruned and a piece breaks off, it may not do much damage. However, if a tree next to the house is incorrectly pruned and a large branch falls from 50 feet (about 15 metres), it can be deadly.

PRUNING LANDSCAPE AND AMENITY TREES

Branch structure and how they are attached to each other in trees falls into 3 categories. Collared unions, collarless unions and codominant unions. Each specific attachment has its own unique way of being cut so that the branch has less chance of regrowth from the cut area and best chance of sealing over and compartmentalising decay. This means that there are 3 types of cuts made, whether that be to remove a little branch coming of another or cutting a whole branch off back to the trunk. This term is often referred to by arborists as "target cutting". Some of the terms used predominantly by arborists and what they entail:

Dead Wooding

Branches die off for a number of reasons ranging from light deficiency, pests and disease damage to root damage. A dead branch will at some point decay back to the parent stem causing abscission and fall off. This is normally a slow process but can be shortened by high winds and extremities of temperature. Therefore the main reason deadwooding is carried out is safety. The situations that usually demand such removal would normally be trees that overhand public roads, houses, public areas and gardens. Trees that are located in wooded areas are usually considered and assessed as lower risk but such assessments would need to consider the amount of visitors. Usually, trees adjacent to the footpaths and access roads are considered for deadwood removal. Another reason for deadwooding is amenity value, i.e. a tree with a large amount of dead throughout the crown looks more aesthetically pleasing with the deadwood removed. The physical practice of deadwooding can be carried out most of the year although preferably not when the tree is coming into leaf. The deadwooding process will speed up the natural abscission process the tree follows. It will help remove unwanted weight; wind resistance the tree carries and can help the overall balance.

Crown - Canopy Thinning

Increase light and reduce wind resistance by selective removal of branches throughout the canopy of the tree. This is a common practice which improves the tree's strength against adverse weather conditions as the wind can pass through the tree resulting in less "load" being placed on the tree. Generally performed on trees that do not have a dense impenetrable canopy as opening a 100% dense canopy up with holes for wind to enter can result in broken branches and uprooting.

Crown Canopy Lifting

Crown lifting involves the removal of the lower branches to a given height. The height is achieved by the removal of whole branches or removing the parts of branches which extend below the desired height. The branches are normally not lifted to more than one third of the tree's total height. Crown lifting is done for access; these being pedestrian, vehicle or space for buildings and street furniture. Lifting the crown will allow traffic and pedestrians to pass underneath safely. This pruning technique is usually used in the urban environment as it is for public safety and aesthetics rather than tree form and timber value. Crown lifting introduces light to the lower part of the trunk; this, in some species can encourage epicormic growth from dormant buds. To reduce this sometimes smaller branches are left on the lower part of the trunk. Excessive removal of the lower branches can displace the canopy weight, this will make the tree top heavy, therefore adding stress to the tree. When a branch is removed from the trunk, it creates a large wound. This wound is susceptible to disease and decay, and could lead to reduced trunk stability. Therefore much time and consideration must be taken when choosing the height the crown is to be lifted to. This would be an inappropriate operation if the tree species' form was of a shrubby nature. This would therefore remove most of the foliage and would also largely unbalance the tree. This procedure should not be carried out if the tree is in decline, poor health or dead, dying or dangerous (DDD) as the operation will remove some of the photosynthetic area the tree uses. This will increase the decline rate of the tree and could lead to death. If the tree is of great importance to an area or town, (i.e. veteran or ancient) then an alternative solution to crown lifting would be to move the target or object so it is not in range. For example, diverting a footpath around a tree's drip line so the crown lift is not needed. Another solution would be to prop up or cable-brace the low hanging branch. This is a non-invasive solution which in some situations can work out more economically and environmentally friendly.

Pollarding

A regular form of pruning where certain deciduous species are pruned back to pollard heads every year in the dormant period. This practice is commenced on juvenile trees so they can adapt to the harshness of the practice.

TYPES OF PRUNING

Regardless of the various names used for types of pruning, there are only two basic cuts: One cuts back to an intermediate point, called heading back cut, and the other cuts back to some point of origin, called thinning out cut.

Removing a portion of a growing stem down to a set of desirable buds or side-branching stems. This is commonly performed in well trained plants for a variety of reasons, for example to stimulate growth of flowers, fruit or branches, as a preventative measure to wind and snow damage on long stems and branches, and finally to encourage growth of the stems in a desirable direction. Also commonly known as heading-back.

Thinning

A more drastic form of pruning, a thinning out cut is the removal of an entire shoot, limb, or branch at its point of origin. This is usually employed to revitalize a plant by removing over-mature, weak, problematic, and excessive growths. When performed correctly, thinning encourages the formation of new growth that will more readily bear fruit and flowers. This is a common technique in pruning roses and for implifying and "opening-up" the branching of neglected trees, or for renewing shrubs with multiple branches.

Topping

Topping is a very severe form of pruning which involves removing all branches and growths down to a few large branches or to the trunk of the tree. When performed correctly it is used on very young trees, and can be used to begin training younger trees for pollarding or for trellising to form an espalier.

In orchards, fruit trees are often lopped to encourage regrowth and to maintain a smaller tree for ease of picking fruit. The pruning regime in orchards is more planned and the productivity of each tree is an important factor.

Deadheading is the act of removing spent flowers or flowerheads for aesthetics, to prolong bloom for up to several weeks or promote rebloom, or to prevent seeding.

PRUNING TECHNIQUES

The general rule to pruning is to always cut in a location where growth will occur, whether the cut is next to a bud or another branch. Cutting a branch beyond where growth will occur will prevent the plant from forming a callus over the cut surface, which in turn will invite insects and infection. It effectively kills all portions of that branch back to the closest branch, bud, or dormant bud clusters, leaving a stub of dead wood. The withered stub will eventually rot away and fall off. All cuts should be relatively smooth since this will aid in healing.

Also, the pruning cut should not be too large when compared to the growing point. For instance, a large cut on a 20 cm trunk down to a 15 cm branch should be fine, but the same cut to the trunk down to a 1 cm twig or bud is considerably less ideal and should be avoided if possible.

Pruning to Bud

A correct pruning cut will allow for quick healing and promote vigorous growth from the closest bud to the cut. The cut should be close enough to the bud to reduce the size of the stub of dead wood that will form from the cut, but far enough away to prevent the bud from being adversely affected by the cut though desiccation. Cutting too close to the bud (under-cutting) sometimes results in the death of the bud, which results in a scenario similar to cutting too far away from the bud (over-cutting). In general, a correct cut should be angled at a moderate 35-45 degree slant such that its lowest point is situated on the same level as the tip of the growth bud. This technique is usually applied when pinching or when cutting-back.

Pruning to a Main Branch

The pruning cut should occur slightly away from and follow the branch collar. When cutting away branches growing directly from the roots, the cut should be flush and level to the ground. This technique is usually applied when thinning or to remove larger dead or damaged branches. When using pruning shears or loppers to remove a branch back to a main branch, the "hook" portion of the shears should always face away from the main branch. This ensures that the blade will not leave a protruding stub and the hook will not damage the branch collar or parts of the main branch.

Large Heavy Branches

Depending on the weight of the branch, the first cut should be a notch on the underside of the branch about a third to half of the way through. The bulk of the branch should then be removed with a follow-through cut slightly above the first cut, thus leaving a limb stub. The purpose of this is to stop the weight of the branch from tearing the bark of the tree from the underside, which would normally occur if the removal was done with one cut. The limb stub ensures that any cracking of the wood resulting from the branch separation is limited to the portion of the wood to be removed. The branch collar should then be located, and can be identified by the strip of rough bark running down from the topside of the branch at its junction with the stem. The cut for removing the limb stub should be just outside the branch collar, leaving a small bump. The bump and the branch collar should not be removed since this action can reduce healing time, which could result in a major infection.

TIME PERIOD

Pruning small branches can be done at any time of year. Large branches, with more than 5-10% of the plant's crown, can be pruned either during dormancy in winter, or, for species where winter frost can harm a recently-pruned plant, in mid summer just after flowering. Autumn should be avoided, as the spores of disease and decay fungi are abundant at this time of year.

Some woody plants that tend to bleed profusely from cuts, such as maples, or which callous over slowly, such as magnolias, are better pruned in summer or at the onset of dormancy instead. Woody plants that flower early in the season, on spurs that form on wood that has matured the year before, such as apples, should be pruned right after flowering, as later pruning will sacrifice flowers the following season. Forsythia, azaleas and lilacs all fall into this category.

Timing of pruning is based on the flowering, fruiting, or growth habits of a plant, its tendency to "bleed," and the fact that pruning usually stimulates a flush of re-growth. Most plants can be pruned at almost any time of year without jeopardising basic survival. However, it is preferable to prune specific plants at specific points in the year.

Pruning According to Season of Bloom

Trees and shrubs that flower before the end of June should be pruned immediately after flowering. Flower buds develop during the previous season's growth, thus, the flowers for the current year's bloom developed last year and overwintered in the bud. If pruned before spring flowering, the flower buds will be removed, thus eliminating flowering. Table 1 shows examples of plants which should be pruned after flowering.

Table 1. Spring-flowering trees and shrubs which should be pruned after flowering.

Scientific name	*Common name*
Amelanchier	Shadblow
Berberis	Barberry
Calycanthus	Sweetshrub
Caragana	Peashrub
Celastrus	Bittersweet
Cercis	Redbud
Chaenomeles	Flowering quince
Chionanthus	Fringetree
Cornus florida	Flowering dogwood
Cornus kousa	Kousa dogwood
Cornus mas	Cornelian cherry
Cotinus coggygria	Smoketree
Cotoneaster	Cotoneaster
Crataegus	Hawthorn
Deutzia	Deutzia
Euonymus	Winged spindle tree
Forsythia	Forsythia
Kalmia latifolia	Mountain laurel
Kolkwitzia amabilis	Beautybush
Laburnum	Laburnum
Ligustrum	Privet
Lindera	Spicebush
Lonicera	Honeysuckle
Magnolia	Magnolia
Malus	Crabapple
Philadelphus	Mock orange
Pieris	Andromeda
Prunus	Flowering cherry and plum
Pyracantha	Firethorn
Rhododendron	Rhododendron and Azalea
Rhodotypos scandens	Black jetbead
Ribes	Currant
Rosa	Climbers and shrub roses
Sorbus	Mountain ash
Spiraea thunbergii	Thunberg spirea

Spiraea x vanhouttei	Vanhoutte spirea
Styrax japonica	Japanese snowball
Syringa	Lilac
Viburnum	Viburnum
Weigela	Weigela
Wisteria	Wisteria

Other trees and shrubs, those which flower after the end of June, should be pruned in winter or early spring before new growth starts. These plants develop flower buds during the spring of the flowering season. Examples of plants of this type are shown in Table 2. Certain plants may be lightly pruned both before and after flowering. This often increases flower and fruit production, and several may produce a second bloom during the year.

Table 2. Summer-flowering trees and shrubs which should be pruned before spring growth begins.

Scientific name	*Common name*
Abelia x grandiflora	Glossy abelia
Acanthopanax	Aralia
Albizia julibrissin	Silk tree
Buddleia davidii	Butterflybush
Callicarpa	Beautyberry
Hibiscus syriacus	Shrub-althea
Hydrangea arborescens	Smooth hydrangea
Hydrangea paniculata	Pee Gee hydrangea
Hypericum	St. Johnswort
Koelreuteria paniculata	Goldenrain tree
Magnolia virginiana	Sweet bay
Rhus	Sumac
Rosa cvs.	Hybrid tea roses
Sorbaria	False-spirea
Stewartia	Stewartia
Symphoricarpos	Snowberry, Coralberry

In any of the foregoing cases, the timing of pruning is based on common sense to maximise flowering of a plant which was planted for its flowers. If your pruning is timed such that flowering is sacrificed, it will not be detrimental to the plant's survival. It will simply mean a loss of one season's floral display. Shrubs or trees that are prized for their fruit should be pruned after the fruit drops or is eaten by wildlife. Although they may flower early in the season, the fruit should be allowed to develop. After the fruit has lost its appeal, then prune.

Pruning in Anticipation of Growth Stimulation

In general, except for the cases already cited, the best time to prune is when the plant will recover the fastest. Severe pruning should be done just before regrowth starts in the spring so bare stubs will be hidden quickly. Pruning in late summer should be avoided since it stimulates succulent growth which may not harden sufficiently to avoid winter dieback. Storm-damaged plants should be pruned as soon after the damage occurs as possible.

BASIC PROCEDURES IN PRUNING

There are three relatively simple techniques basic to all pruning situations. Pinching is usually done by hand, and this is a good way to control plant size. (Fig. 1.) Thinning completely removes some branches back to a main branch, trunk, or soil line. Do not cut into the branch collar when making a thinning cut back to a trunk or main branch; that is, do not cut so near the trunk that you cut through the area at the base of the limb adjacent to the main trunk, known as the branch collar. Such a cut allows for infection to spread into the part of the plant you wish to keep. Cut only the branch to be removed, about ½″-2″ from the main trunk (depending on age). (Fig. 2.)

Figure 1. Pinching out Growing Tips of Shoots

Heading back involves shortening branches back to a good bud or lateral branch. A proper heading back cut should not leave a stub. Make your cut about ¼″ above an active bud or lateral branch.

Figure 2. Thinning of Limbs back to a Main Trunk

Wound Dressing

Wound dressing or tree paint is not essential; recent research has pointed out that tree paint or wound dressing is not as advantageous as previously thought. Dressings may actually harbor disease organisms rather than exclude them. It has also been determined that wound dressing slows the wound callusing process, rather than speeding it up. The only reason for painting a pruning cut is a cosmetic one. When appearance is a factor, the painting should be done with latex paint. A good, clean unpainted pruning cut, while perhaps unsightly at first, will probably callus faster than a painted one. On the basis of tree health alone, pruning cuts should not be painted.

Basic Safety Rules for Pruning

1. Call in a professional for large trees or for jobs you don't have the equipment for.
2. Keep all equipment sharp and in good repair.
3. Use equipment only for the job it was designed to do.
4. Be conscious of electric lines when pruning near them.
5. If a power line is touching a tree limb, call the power company fast and stay clear of the tree.
6. Never climb a tree without a safety rope, with or without a ladder.
7. Keep your fingers clear when using hand clippers.
8. Use care in handling pruned limbs and brush to avoid eye injury.

PRUNING OF SPECIFIC TREES AND SHRUBS

Deciduous Shrubs and Trees

Always allow a shrub or tree to develop its natural shape, except when special effects are desirable, such as for hedges. "Haircut" pruning should be avoided. Perfectly sheared globes and squares make a mockery of the plant's natural form and beauty. (Fig. 3.) Instead, use the thinning technique on both shrubs and trees. Cut the branches at different lengths, ¼" above an active bud. Remove twigs or branches selectively and thus reduce overcrowding. Some stems should also be removed at ground level. The length of new shoots should be reduced 1/3 to ½ of their length, which induces side shoots to develop.

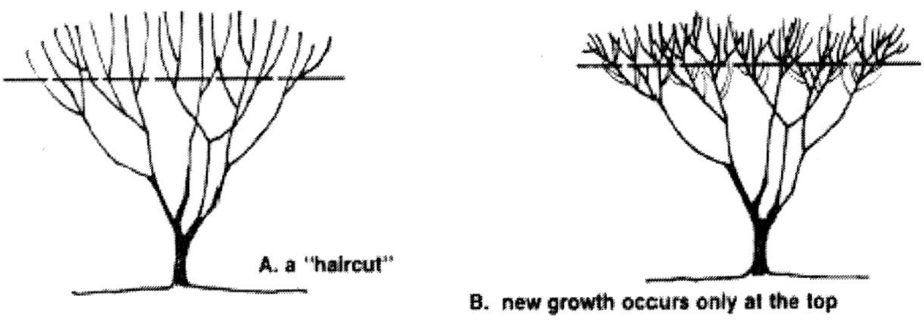

Figure 3. "Haircut" Pruning should be avoided

Cutting above a bud prevents dieback of the stem and encourages a new branch to develop from the bud. The haircut technique causes a dense growth at the ends of the pruned branches which shades the rest of the plant, thus causing the plant to eventually develop a leggy appearance. If a shrub develops a weak, dense growth, thin out many of the smaller branches and twigs. This promotes the vigorous growth of the remaining branches. Also remove branches which tend to rub against one another, opening wounds for the entrance of disease.

Always remove dead, damaged, or diseased branches and limbs from established plants as well as newly planted trees and shrubs. The plant should then develop new, stronger growth, free of diseases. Also, remove dead flower branches, dead flowers, and old fruit stocks as soon as the flowers have wilted or the fruit has dropped. This stimulates new growth and helps to make a stronger tree or shrub. It also encourages plants such as rhododendrons to produce more flower buds for next season.

Evergreen Shrubs and Trees

With evergreen shrubs, avoid shearing to artificial shapes. If the evergreen plant has a soft, feathery appearance, do not cut it square or make a round shrub out of it. Prune using the thinning technique, thus keeping the natural shape of the evergreen. Remember, the non-green portion of needle-leaved evergreen branches does not normally put out new branches. Therefore do not cut branches back to the old wood. Reduce new growth annually, and when removing the larger branches for thinning, cut close to the main trunk, leaving no stubs. Heavy thinning is needed only every few years.With certain evergreens such as yews, a new flush of growth will occur in the early fall. Head back these long shoots to keep the plants in shape. Broadleafed evergreen shrubs are pruned in the same manner as narrowleafed evergreens, by thinning and heading back. Light pruning every year is preferred, but heavy pruning every three years is acceptable. Rhododendron species benefit from removal of flower heads immediately after flowering. Most types, such as hollies, pyracantha, azaleas, and euonymus, can be cut back severely, but avoid cutting all the way to the ground. Holly trees may be pruned at Christmas time. To thicken the new growth of coniferous trees such as pines, spruce, or fir, pinch out ½ of the candle when it is approximately 2" long in the spring. Do not use shears, since they damage the needles that are around the candle and cause the cut edges to turn brown. This gives the tree an unsightly appearance. Do not top or remove the central leader, if the natural growth habit of the tree is desired. (Fig. 4.)

Figure 4. Pinch pine candles when new growth is about 2 inches long.

If the terminal of a pine or spruce has been lost, it is necessary to aid the plant in growing a new terminal shoot. Without assistance, a single terminal will probably not be re-established, and multiple leaders will result. To form a new terminal, bend one of the youngest lateral branches near the terminal into an upright position.

Secure it to the dead terminal stub or insert a stake for rigidity. After a season, this branch will take over as a terminal shoot. Once this occurs, remove the stub or stake, and the plant will resume its characteristic growth habit. (Fig. 5A and B.)

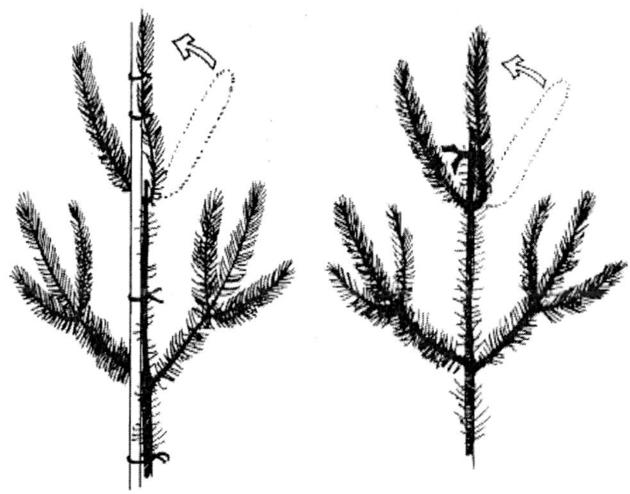

(a) *(b)*

Figure 5: (a) Replace a lost terminal leader by tying a lateral branch in a vertical position and securing it to the stub of the dead terminal. (b) The Lateral Branch can also be Secured with a Stake

Evergreen trees have leaves that persist year round, and include most conifers and some broad-leaved trees. Evergreen trees generally need less pruning than deciduous trees. Conifers are distinguished from other plants by their needle or scale-like leaves, and their seed-bearing cones. Because conifers have dominant leaders, young trees rarely require training-type pruning. The leader is the vertical stem at the top of the trunk. If a young tree has two leaders, prune one out to prevent multiple leader development. Selective branch removal is generally unnecessary as evergreens tend to have wide angles of attachment to the trunk.

Evergreens are grouped on the basis of their branch arrangement. Pines, spruces, and firs have whorled branches that form a circular pattern around the growing tip. The annual growth of a whorl- branched conifer is determined by the number of shoots that are pre-formed in the buds. Whorl-branched conifers usually have only one flush of growth each year in which these pre-formed shoots expand into stems that form the next whorl. The second group of evergreens are those with a random branching habit. Yew, arborvitae, cedar, false cypress, and juniper are all random-branched species.

Corrective pruning for evergreen trees consists mainly of dead, diseased, or damaged branch removal. Remove dead wood promptly, by cutting dead branches back to healthy branches. When pruning diseased branches, make thinning cuts into healthy wood, well below the infected area. Thinning cuts remove branches to their points of origin or attachment. Disinfect tools between each cut with products such as "Lysol," "Listerine," or rubbing alcohol. Tests have shown that "Pine-Sol" and household bleach are highly corrosive to metal tools.

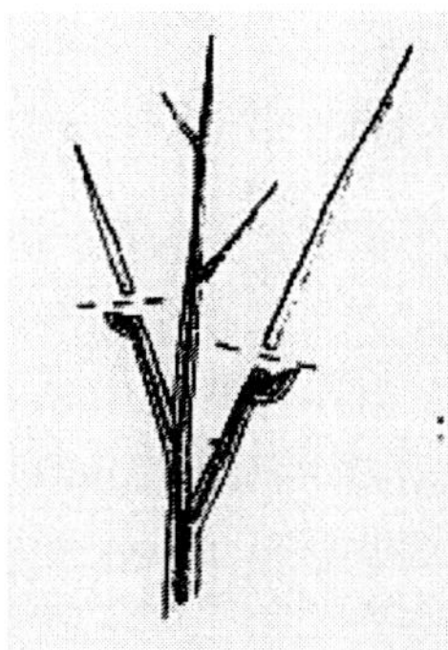

Figure 6. Pruning back multiple leaders

Allow evergreen trees to grow in their natural form. Don't prune into the inactive center (no needles or leaves attached) of whorl-branched conifers because new branches won't form to conceal the stubs.

When a tree's leader is lost due to storm damage or disease, replace it by splinting to a vertical position the upper lateral on the highest branch. Prune all laterals immediately below the new leader. Use wood or flexible wire splints, removing them after one growing season.

Current pruning recommendations advise against pruning branches flush to the trunk. Flush cutting is harmful in several ways: it damages bark as pruning tools rub against the trunk, it removes the branch collar, and it goes behind the branch bark ridge. The branch collar is the swollen area of trunk tissue that forms around the base of a branch. The branch bark ridge is a line of rough bark running from the branch-trunk crotch into the trunk bark, less prominent on some trees than on others.

The best pruning cut is made outside the branch collar, at a 45 to 60 degree angle to the branch bark ridge. Leave the branch collar intact to help prevent decay from entering the trunk.

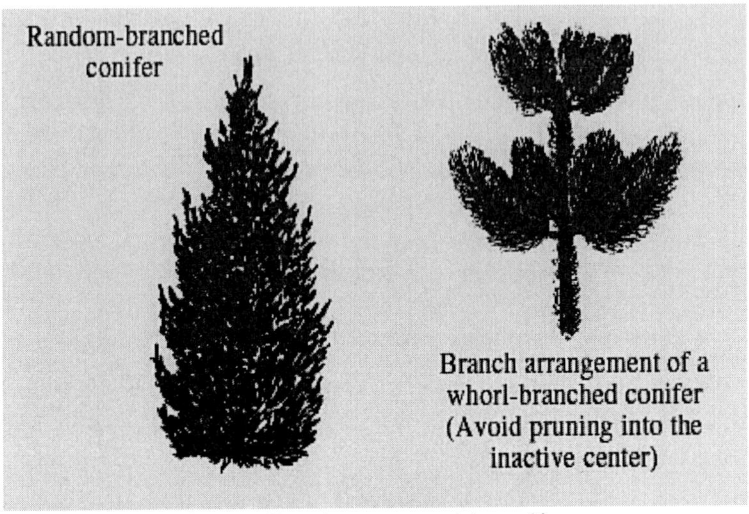

Figure 7. Branch arrangement in conifers

Whenever removing limbs greater than one inch in diameter, use the three-cut method to avoid tearing bark. First, about 12 inches from the trunk, cut halfway through the limb from the underside. Second, about 1 inch past the first cut, cut through the limb from the top side. The limb's weight will cause it to break between the two cuts. Make the third cut outside the branch collar.

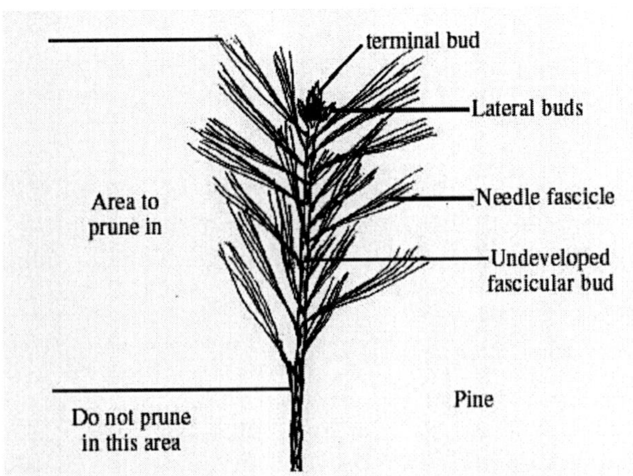

Figure 8. Areas to prun

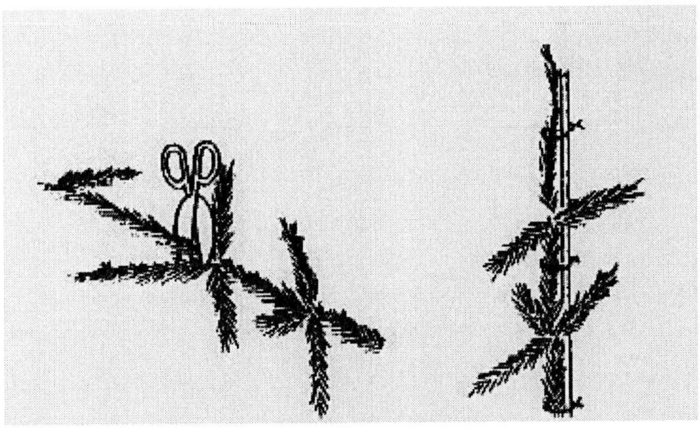

Figure 9. Use splint to train a replacement leader

Don't coat pruning cuts with tree paint or wound dressing, except for control of certain disease-carrying insects. These materials won't prevent decay or promote wound closure. Some tests, however, have shown wound dressings to be beneficial on trees that are susceptible to canker or systemic disease.

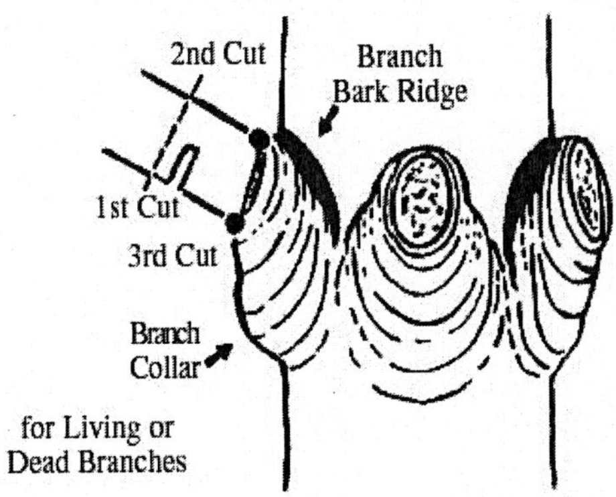

Figure 10. Branch bark ridge and branch collars in conifers

Pines and other whorl-branched conifers become denser if new growing tips ("candles") are pinched in half as they expand in the spring. Pinch by hand, as pruning shears will cut the expanding needles and leave them with brown tips. Most evergreen pruning is done for corrective reasons, so seasonal timing is not as important as it is for deciduous species. Pruning during dormancy is the most common practice and will result in a vigorous burst of spring growth. Whenever unexpected damage from vandalism or bad weather occurs, prune immediately.

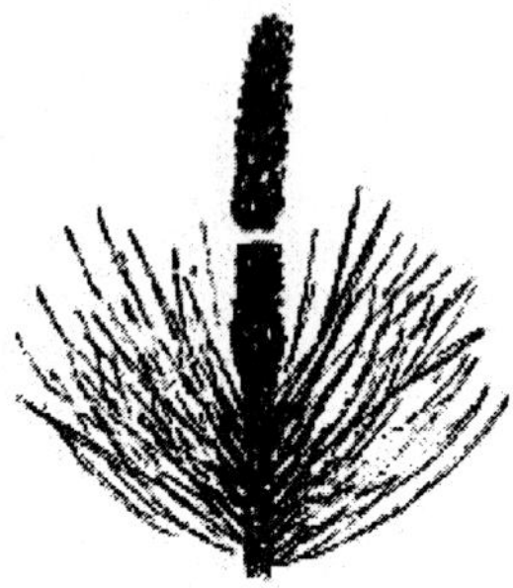

Figure 11. Pinc back pine candles

There are, however, certain evergreen pruning activities for specific times of the year. Prune random-branched conifers in early spring when new growth will cover the pruning wounds. "Candles" of whorl-branched conifers must be pinched back in mid to late spring. Maintenance pruning of random-branched conifers is done in summer to keep plants within a desired size range. Remove spent flowers of evergreen magnolias at the end of their blooming season to stimulate new growth and development of a thicker crown. During the Christmas season, minor pruning for decorative purposes usually causes no harm.

Newly Transplanted Trees

In the digging process at transplanting time, bare root trees and shrubs suffer root damage as well as a loss of roots. To compensate for this loss of roots, the leaf area of the plant should be reduced by 1/3 to reduce water loss due to transpiration and evaporation. The natural form of newly planted trees and shrubs should be preserved by thinning. Remove branches and parts of branches by cutting at different lengths as previously described. (Fig. 12.) When the tree is approximately 2 years old, establish good branch spacing (Fig. 13). Branches should arise alternately from the main stem and be well-spaced radially. This will allow stronger limbs to develop. Select branches with the widest angles in the crotch. The wider this angle, the stronger the limb attachment. On most shade trees, the top-most growing point is critical in achieving a form typical of the species. This growing point is called the central leader, and there should be only one. It should not be cut.

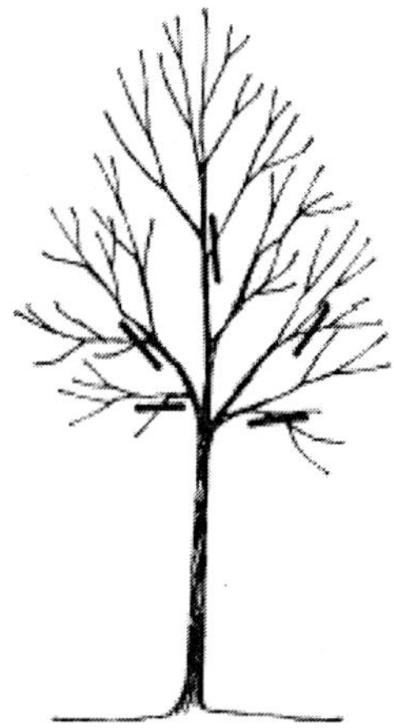

Figure 12. Use the thinning technique to remove 1/3 of the leaf area of newly-planted trees

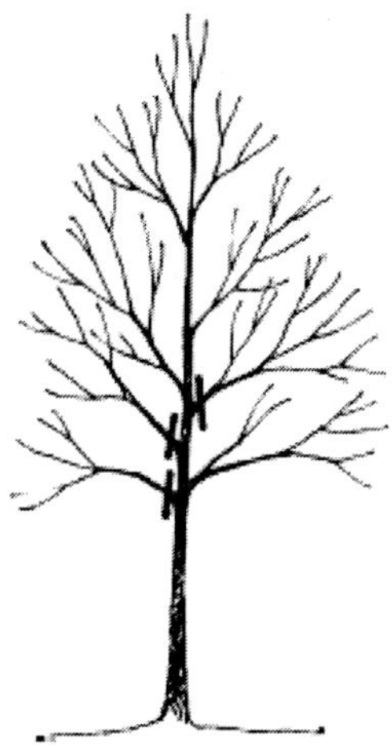

Figure 13. Establish Alternate Branching

Large Trees

Proper pruning also helps to prevent injury and damage to life and property. This usually involves the removal of large branches or limbs from trees. Where tree limbs are near power lines, call the power company and ask them to remove them. Do not try to remove tree limbs from power lines yourself.

Low-hanging branches may cause injury to individuals mowing the lawn or walking on the street. Also, branches sometimes rub against the house and roof. To remove the branches that are over 1" in diameter, use the double cut method. If the double cut method is not used, the branch will tear and splinter the trunk. This removes a large portion of the bark, causing a large wound which calluses with difficulty and may further result in permanent damage to the tree. (Fig. 14)

Figure 14. Failure to use the double cut method can result in trunk damage and bark stripping

To double cut, first cut halfway on the underside of the limb. Then make a second cut through on the upper part of the limb. When the branch is removed, there is no splintering of the main tree trunk. Then remove the stub by conventional methods, taking care not to cut into the collar.

Beware of rejuvenation techniques sometimes used on large old trees. Homeowners are sometimes "conned" into having the tops of old trees completely cut back, leaving only the stubs. These stubs eventually decay. Also, since the tree is in such a weakened condition, it may die prematurely. If you have large limbs that need to be removed, secure the professional services of an arborist. An arborist can drop work a tree, that is, lower the height, by removing limbs that are causing problems. When completed, the tree will still retain the beauty of its natural shape. Remember, a tree that has a trunk diameter of 20-24 inches may be worth up to $10,000 on a replacement value scale. Therefore, if you have a large, valuable tree, secure the services of a professional arborist.

Old, Overgrown Shrubs

Another aspect of pruning is the renewal of declining shrubs. In renewal pruning, remove one-third of the old, mature stems per season. These large, old branches are removed at the ground level. Leave the young vigorous branches. The water sprouts that develop should be cut back to different lengths and encouraged to develop into strong branches for the shrubs by the thinning process.

Lilac wood often is more than 3 years old before it flowers. Therefore, large, overgrown lilac bushes can gradually be cut back over a period of years, but do not remove all the old flowering wood until the new growth begins to flower. Then the bush will flower every year and will not have any barren years. If it is essential to immediately reduce the size of a shrub, rejuvenation pruning is appropriate for some species (Table 4). Cut back the entire top of the plant to the ground line. Many new shoots will grow from the base, and they will require thinning. As much as ¾ of the new growth should be removed, depending on the species.

Table 4. Landscape plants which may be rejuvenation pruned (completely cut back to the ground).

Scientific name	*Common name*
Buddleia davidii	Orange-eye
Forsythia	Forsythia
Hibiscus syriacus	Shrub-althea
Hydrangea arborescens	'Grandiflora' Hills-of-snow
Hydrangea quercifolia	Oakleaf hydrangea
Ligustrum vulgare	Privet
Lonicera	Honeysuckle
Spiraea	Spirea
Syringa	Lilac

Some plants frequently suffer winter die back of all above-ground stems. These may be safely rejuvenation pruned to produce rapid new growth. These plants are considered to be usefully winter hardy. They can be used for landscaping purposes similar to an herbaceous perennial, even though severe rejuvenation pruning often is necessary.

Hedges, Espaliers, and Topiary

Plants are occasionally sheared to unnatural shapes. Hedges as barriers are the most common example. However, hedges are often improperly pruned, with the top trimmed flat and the sides sloped inward so that the base is more narrow than the top. This shades the lower portion of the hedge, preventing dense leaf growth. Thus, the plants become leggy, the hedge becomes straggly, and the screening function is not achieved. In proper hedge shearing, the top is shaped more narrow than the base. This means that light can penetrate to the lower portions of the plants, growth can be maintained, and a full appearance over the entire height of the hedge can be attained. This simple technique helps the hedge remain a satisfactory barrier for a long period of time.

Two other types of pruning plants to unusual shapes are espalier and topiary. Both originated in European gardens and are very time consuming. An attractive plant requires both dedicated effort and constant attention. Not all landscape designs are appropriate for such plants, and in all cases, their use should be limited to focal points due to their highly unusual appearance.

Espaliering is the practice of training a tree or shrub to grow flat. Almost any tree or shrub can be trained flat by continually removing growing points that go in unwanted directions. Allow the rest of the growing points to develop in their own way. Before locating a plant next to your house for espaliering, place an iron or wooden support a few inches away from the house. This prevents any disfiguration of the wall and allows for plant support and easy plant removal at a future date.

Topiary is a shearing technique occasionally done on boxwood, juniper, pyracantha, yew, and privet. Part the branches to find a basic form. Remove limbs you don't want. Wherever you want to form a clump, leave some branches and shear them to the outline you desire. Remove all twigs and leaves along branches between clumps. Each season new branches and leaves will tend to fill in the spaces between clumps.

REFERENCES

Arbury, Jim. et al. 1997, *The complete book of plant propagation,* Imprint: London: Reed International Books, 1997.

Barrow, Ron. 1994. *A student handbook for the study of plant propagation - for the Advanced Certificate in Horticulture.* Published Kew, Vic. Adland Horticultural Consultants.

Bird, R., 1993. *The propagation of hardy perennials,* B.T.Batsford, London.

Brickell, Christopher 999. *Plant propagation,* The Royal Horticultural Society. 1999.

Bryant, Geoff. 2002. *Plant propagation A-Z : growing plants for free,* Lothian.

David, Tim D, Haissig, Bruce E, Sankhla, Narendra. 1998. *Adventitious root formation in cuttings.* Published Portland, Or. : Dioscorides Press.

Edmanson Jane. 1991. *Cheap and easy propagation.* Published Melbourne Lothian.

Gardiner, A. 1998. *Modern plant propagation.* Published Port Melbourne Lothian.

Hawthorne, 2003. *Complete book of plant propagation,* Reed Books.

8

Vegetable Gardening

Gardens have been established next to homes since prehistoric times. The most important characteristics of home gardens are their location adjacent to homes, close association with family activities and a wide diversity of crop and livestock species to meet family needs. They have played a central role in household security for food, fuel, fibre, materials and even land ownership, as people changed from an exclusively hunting and gathering lifestyle and settled in small communities. Small-scale farming worldwide typically combines production of different crops, vegetables and livestock.

Diversity in size, form and function make it difficult to define home gardens, but their place in the farming systems of the rural landscape is readily recognised. Cropping and grazing areas surround a settlement; there may be large-scale monocultures such as wheat or sugar cane; further away there may be forest or other common land used periodically for grazing, hunting and gathering firewood, materials and seasonal forest foods. In the village, the small area surrounding a house provides good conditions for a garden.

The village is usually close to a source of water and is usually better protected from floods and foraging animals than other farmland. There is good access by path or road, and it is the family's central living area. A home garden can be defined as a farming system that combines physical, social and economic functions on the area of land around the family home. The area is used as a place of work and for storage and processing of farm produce; it is also a place where people live and dispose of wastes. While the focus of this book is on the home garden's economic functions—production of crops, livestock and other items—the other functions of a home garden are often equally important. The diversity of household needs is reflected in home gardens which can include staple foods, fruits, vegetables, materials, condiments, stimulants and medicines.

Home gardens are developed for some or all of the following purposes.

- To supply nutritious food and some staple foods all year round including food plants, fish, poultry and snack foods such as fruit and special food reserve resources such as root crops, trees and livestock.
- To generate income from the sale of garden produce; sales and value-adding can contribute substantially to a family's income.
- To support important farm-development activities; some farm inputs come from home garden activities such as plant propagation, raising and housing draught animals and making and repairing tools. New crops and farming techniques are often tried out in the home garden, which is also an area for drying, processing and storing farm products.
- To provide a healthy, comfortable and beautiful environment. A productive home garden can contribute to safe recycling and management of household wastes through com-posting or as animal feed, by utilising waste water and by providing a place for a latrine. Gardens offer privacy, shade and flowers for a family and their visitors to enjoy.

SYSTEMS, RESOURCES AND CHOICES

Home gardens worldwide are integrated into family life. They may be divided broadly into traditional gardens, resulting from a long history of adaptation of plants to local needs and conditions, and model gardens, often developed with external support, ideas and imported technologies. In urban areas and isolated rural areas, a traditional kitchen garden may be inexpensively established—a small plot from which vegetables and garnishes are taken each day to improve meals. It can be watered with the wastewater from dishwashing and bathing.

Depending on local environmental sensitivity and available recycling technologies such as composting and biogas production, nutrients can be utilised from kitchen and human wastes. Requiring a little more space and capital invested, traditional mixed gardens integrating poultry, other livestock and fish ponds provide productive opportunities for waste transformation and nutrient recycling. Agroforestry gardens maximise use of scarce land by cultivating crops in multiple layers -trees, vines, understorey and root crops. Living space, boundaries and materials are integrated into gardens. Trees provide shade and shelter under their canopy and their roots stabilise soil around the home.

Multipurpose plants, such as sea buckthorn in cold areas, *Leucaena glauca* in tropical areas and even cassava, are planted as living fences to provide crop protection, privacy, firewood, materials, food and animal fodder. In parts of Asia, parents traditionally plant a neem tree (*Melia azadirach*) in their home garden for every child born in their family, so that when they are adults there is timber for them to build their own house.

Oil from leaves and seeds of the neem is a natural pesticide and is now planted in African home gardens. Made from what it grows, the traditional rugo in central Africa is a cluster of round thatched huts fenced by grasses, wood and crop terraces in concentric circles to control erosion and afford protection from wild animals. It is a home for the extended family and there are huts for livestock, food stores and even compost. Most communities have enough people to allow viable specialisation. Market gardens are specialised to meet market opportunities such as daily fresh vegetables or seasonal fruits.

Dairy products, eggs and other livestock products have a fairly stable market demand even in small communities. Nurseries propagate and sell trees for gardens, plantations and woodlots, or vegetable seedlings to meet local demand for seasonal field cropping. Floriculture, potted plants and ornamental plants are another specialised market-oriented garden system, particularly in peri-urban areas. A household makes choices on the size and nature of the home garden system and its purposes, depending on their needs, resources and market opportunities.

GARDENS—PLACES FOR INNOVATION

Like all forms of agriculture, home gardening today involves environmental modification, ranging from intensive systems such as greenhouses, where all aspects of plant life cycle are managed, to extensive systems such as fruit orchards, which may receive little care after planting.

Home gardens are places where innovation is nurtured; technical and market developments have benefited home gardens. Innovation and trade have led to many products being adapted in home gardens, some subsequently for cropping in larger fields. Breeders of plants and livestock have for centuries selected lines for small-scale production; access to this diversity of genetic resources improves productivity and choices.

Technology such as inexpensive poly-thene film can be used to improve growing conditions. Products and practices can be selected to suit changes in household needs and market opportunities. Agrochemicals have a place in agriculture, but organic production practices are favoured by the proximity of gardens to homes, community environmental concerns and opportunities to have fresh food available daily. Recent increases in market demand for organically-grown foods have led to innovations in production technology suitable for home consumption and small entrepreneurs.

TRENDS IN GARDENING

Climate, local food preferences and trade influence regional differences in home gardens. The naturally high biodiversity of certain tropical regions, combined with long exposure to trade routes, has led to high diversity of plants and animals and the existence of fish ponds in many gardens.

In Java, Indonesia and Kandy, Sri Lanka, traditional gardens—loosely managed, multilayer agro-forestry gardens- use 10 or more different species to produce food from below ground upwards: root crops, leaf vegetables, climbing vines, low trees and emerging canopy trees. These agroforestry gardens are common where competition for land is high. Such diversity of fresh food disappears when changes in economic policy, employment patterns and increasing population affect traditional land use.

In parts of China, the pressures of high population and limited availability of land and nutrient resources have forced innovation and intensity in closely managed gardening systems. In some cases, ten crops can be harvested from a garden bed in one year. Despite this intensity, gardeners have used organic practices so effectively that fields near Chengdu cultivated during the Han dynasty are still fertile after 20 centuries of continuous use.

In other areas, these pressures, coupled with changing market demands and the advent of agro-chemicals without adequate advice and training, have led many farmers to use methods that are not part of their familiar local knowledge. This has resulted in cases of poisoning from pesticide residues, toxic nitrate levels in groundwater and the erosion of soil fertility that has made gardening impossible, particularly in peri-urban areas. Producing a livelihood from the soil—sustainably -employs a sometimes fragile ecological balance.

FUTURE DIRECTIONS

How can home gardens contribute to food supplies, rural employment and incomes in the future? Among other factors, the growing population in many parts of the world has led to increasing poverty and insecure food supplies. It is predicted that by 2010 the world's population will have reached 7.3 billion, over 90 percent of whom will be living in developing countries. Creation of urban employment to keep pace with population increases will be difficult if not impossible; maintaining viable rural employment is therefore vital to prevent an explosion of urban poverty. In *The State of Food Insecurity in the World, 2001*, FAO estimates there are about 815 million undernourished people, 95 percent of whom are in developing countries; over 20 percent are children. Poverty in rural populations, combined with undeveloped transport and food production systems, restricts them largely to locally grown products.

The role and contribution of home gardens in addressing these problems have been recognised by development organisations since the 1970s, largely as a result of research into farming systems that resulted in greater understanding of farmers and their households in agriculture and rural development organisations. In order to improve rural and peri-urban livelihoods in developing countries, development organisations have promoted home gardening with one or more of the following objectives:

- to reduce poverty;
- to diversify income and rural employment;
- to improve the quality and quantity of household food supply and improve nutrition;
- to improve the status of women;
- to improve water and waste management at household and community levels;
- to reduce pressure on wild food resources.

Some specialist research institutions, such as the Asian Vegetable Research and Development Centre, have developed advanced home-garden systems for specific purposes, for example, to provide cost-effective daily Vitamin A requirements. Others, such as FAO, have helped communities worldwide to adapt home garden technologies for a wide range of purposes, including nutrition improvement, school gardens, women's incomes and rural livelihood diversification. These efforts and investments have generally been successful when people have had access to enough resources and services to enable them to make their own choices.

A home garden as small as a few square metres can provide enough Vitamin A, Vitamin C and other nutrients to meet a growing child's needs. It can provide women with a cash income and increase the resilience of families to withstand shocks to their supplies of food, their income and their health, and so avoid a further slide into poverty. There is still great potential for home gardening to improve the livelihoods of people in developing countries, especially in Africa, Latin America and the dry regions of Asia. Successful approaches need only to be spread by local adaptation and limited external support.

Small-scale technologies to protect growing and stored crops from cold have increased the potential for gardening in cool temperate and high-altitude ecological zones in northern and central Asia. The close link between gardens and homes means that home gardening is an important tool for communities to use in keeping pace with socio-economic development; it can be effective as a stand-alone initiative or as a component of a large area-based rural development project.

BENEFITS OF VIABLE HOME GARDENS

A number of benefits from home gardens and the influence of gardening from ancient to recent times have been described in the preceding sections. This section focuses on how home gardens fit into the livelihood system of smallholders today. Viable home gardens improve the ability of smallholders and their communities to meet interrelated concerns of food security, nutrition, health and economic security. Potential benefits and beneficiaries include:

- income and enhanced rural employment through additional or off-season production;
- improved food security;
- increased availability of food and better nutrition through food diversity;
- decreased risk through diversification;
- environmental benefits from recycling water and waste nutrients, controlling shade, dust and erosion, and maintaining or increasing local biodiversity.

Gardens Reduce Food Losses

The location of the garden close to the home reduces the risk of food losses from foraging wild animals and from theft. In the household farming system, most staple foods are usually supplied by one or more fields of a crop. Such fields are typically at a distance from the home, and a family member may have to stay there overnight in a makeshift hut in order to protect it at harvest times. Crops are usually cultivated as a monocrop—a single species all planted at the same time—in order to maximise growth and labour efficiency. This low species diversity increases the risk of loss from disease and pests, which easily multiply and spread under such conditions.

Planting the crop at a single time exposes it to loss by drought and bad weather. In contrast to this, a home garden's high species diversity and staggered planting times increase the likelihood that some crops survive. In the Philippines and the Pacific Islands, patches of taro (*Colocasia* and other related species) grown in gardens ensure that a family has food after a typhoon or tropical cyclone has wiped out other crops. Plants such as the onion family (*Allium* sp.) and neem tree (*Melia azadirach*) are planted in gardens to repel pests. Shade and shelter for the home also protect the home garden crops, which is not the case in open fields. The home garden also provides a secure place to process and store crops.

Gardens can make an important contribution to food security as an additional food source or by supplying off-season production. In all but the coldest and driest climates, vegetables can be planted and harvested for most of the year. Trees often bear fruit or nuts at different times of the year compared to staple food crops. In tropical climates, papaya (*Carica papaya*) and banana (*Musa paradisiaca* L.) may be harvested almost year round.

In subtropical areas, fruits from South America, such as feijoa (*Feijoa sellowiana*) and avocado (*Persea americana*), and from Asia, such as mandarins (*Citrus sinensis*) and jujube (*Ziziphus jujuba*) are harvested from late summer into winter, when few other fruits are available. In dry and temperate areas, relatively non-perishable nuts such as walnut (*Juglans regia*) and cashew (*Anarcardium occidentale*) provide useful food and trade items that can be stored at home or sold in markets. Livestock and aquaculture are integral parts of many home garden systems, whether the climate is humid or dry.

In a home garden, terrestrial and aquatic animal production can generate high levels of output and income and improve family nutrition while contributing to waste management and water and nutrient recycling. In many countries, animals are a low-cost source of high-value food containing protein, fats and micronutrients. In Latin America, guinea pigs—known locally as cuyes -are fed on kitchen food scraps supplemented with fresh fodder; in Asia, snails are a protein source that feed on fish pond weeds and kitchen scraps.

Labour and Time Efficiency

Travel to and from crop fields can be arduous and time-consuming. Working in a home garden can give equally high returns from labour without the need to be far from home. Work in a home garden is generally not as physically demanding as field preparation and crop weeding, because the area is smaller and working conditions are better. Reduced work effort, particularly for women and girls, helps to balance the household workload, most of which is usually done by women.

Daily household routine have shown that although men and women may appear to work an equal number of hours each day, chores such as preparing meals, childcare, cleaning, livestock care and fetching water and firewood are often not fully considered. These tasks most often fall to women and are centred on the home. This publication does not aim to promote home gardening as women's work only; it makes the point that work in the home garden is readily integrated into the daily household chores and can help women to earn an income while doing so. One of the most important roles of women is caring for small children.

During advanced pregnancy, a woman should not do heavy manual field work or endure long days with insufficient regular food, in order not to risk her own or the baby's health. Infants need regular breast-feeding and attention; and after six months, breast milk becomes inadequate and complementary foods need to be provided. Being at home during these months is the best solution: home gardens help to provide profitable but light work, and a mother can look after her own and her child's health and nutrition.

Nutritious food is available from the home garden every day; it can be prepared freshly and safely at home. Feeding an infant will also be easier at home. Other labour efficiencies can be gained from the home garden. Nursery beds can provide advanced seedlings, which when planted in fields require less weeding and take more rapidly than when seeded directly as field crops. Winnowing, drying, milling and other post-harvest processing may be more efficiently done in the home garden. Working in the garden near drying crops ensures that they can be brought in from rain; a gardener can also keep watch for browsing domestic animals. Waste from crops processed in or near the home garden provides feed for livestock and compost for garden fertilisation.

Improved Working and Living Conditions

The ecological association between people and plants in a home garden is an ancient one. People and plants thrive in places that have adequate shade, shelter, light, water and nutrition. This makes for better working conditions for:

- highly intensive enterprises such as poultry and pig raising, nurseries, market gardens or floriculture;
- post-harvest processing or value-adding activities;
- non-farm employment such as clothes making and carpentry.

Home gardens can provide environmentally sound opportunities for waste disposal. Composting is commonly used for household wastes including kitchen waste, paper and other materials. In flood-prone areas such as the Ganges delta in Bangladesh, home gardens can literally anchor the family home. Plants like taro, coconut and thatching palms hold the soil together while inundated. In response to regular devastating floods in the Mekong delta, Vietnamese people have built pontoon-like houses that float during floods. Home garden trees adapted to the irregular floods and rich soils, for example palms and durian, continue to provide protection around the floating house.

Enhanced Social Standing

In all countries, certain groups are vulnerable to food insecurity, poverty and poor social standing. Gardening is often used to reduce a family's vulnerability. In urban areas across the world, food crops are grown on rooftops and balconies, in backyards and community gardens, alongside roads and in vacant lots. They provide fresh, marketable food that supplements family diets and boosts family incomes. Urban and peri-urban agriculture can fill critical food-supply gaps for poor city dwellers, particularly where rural infrastructures and farm-to-market distribution systems are poor.

Disabled and elderly people are often considered non-productive dependents in a household. Limited care of a home garden and other household activities provide them with safe and feasible opportunities to contribute to household food and income.Home garden systems are readily accessible to the poor. While gardening for subsistence is common and valuable to the household, gardens offer a way to generate a small income rapidly. A small investment in seeds and a small amount of labour can provide a return from the sale of vegetables within six to eight weeks.

Limited access to land is often a characteristic of poor families, but need not be a major constraint—a home garden can be established on a small area. Gender inequities increase the vulnerability of women to poverty and malnutrition, make it difficult for them to earn a livelihood—especially single mothers—and reduce their social standing.

Marketing garden produce can be an important source of independent income for women, particularly in households headed by women or where men migrate for long periods, or in cultures where women traditionally feed the family through their own work. In parts of Africa severely affected by HIV/AIDS, home gardens provide food for single-parent or no-parent households.

Traditional divisions of labour and responsibility often mean that men sell major cash crops or staple food crops and control the resulting income. Income from gardens controlled by women allows them to purchase items that are important to improving their social status in families and communities where men have a dominant social position.

Better Skills Training

Acquisition of skills is facilitated by home gardens, because they are close to homes in the community, relatively small in scale, potentially viable with small investment and suitable for a wide range of people. Typical village social interaction means that crop and animal technologies and business management skills and concepts can be readily exchanged. Seeds, cuttings, poultry hatchlings and fish fingerlings are readily and cheaply traded; novices can learn integrated cultivation and husbandry practices from what is achieved in established home gardens. Compared to conventional training, farmer participation offers greater opportunity for social learning, which contributes greatly to generating innovation.

Farmer field schools, for example, give farmers the confidence to work together on low-cost sustainable farming practices. Better skills transfer and acquisition rapidly increases the variety of livelihood improvement options available to small farming families. Group credit schemes using revolving funds have become a popular and fairly successful poverty-reduction tool in rural development.

Groups typically consist of four to six women, who in turn use the US$30-US$50 credit fund loan for small enterprises in home gardens. The necessary business management training is made relevant by using real-life information from group members' home gardens and household budgets. In Asia, about 60 percent choose to raise animals, so an animal-husbandry training session provided in support can easily and effectively reach a number of people.

Added Value to Livelihoods and Trade

Viable home gardens create opportunities for input suppliers, processors, small manufacturing, traders and other service providers, as well as generating income, much of which is spent in the community. Adding value to crops and livestock through processing, storage and small manufacturing increases the livelihood options for rural households. Milk, feathers and fibre from poultry and livestock are commonly worked into saleable items by home-scale processes. Processing perishable products extends shelf life, reduces losses in transport and can enable a regular supply to be maintained to markets. Development of small businesses can multiply the employment and livelihood opportunities for rural communities. Many microenterprises grow into thriving businesses, often so specialised that their home garden origins may no longer be recognisable. Many small-scale street vendors and food-stall holders can make a living for themselves and their families by processing or selling food.

In Malaysia, for example, preparation and sale of street food employs more than 100 000 people and generates more than US$2.2 billion in sales annually. If product quality can be assured, street food provides adequate and inexpensive nourishment for many urban inhabitants. Supply of agricultural inputs is often controlled by large state or private corporations, which may offer only a narrow range of high-volume items such as fertiliser and main crop seeds such as rice and maize. In contrast, viable home gardens require a wide range of high-quality vegetable and herb seeds, grafted fruit trees, young livestock for fattening, materials for pest exclusion and environment modification, and services such as paraveteri-nary services. This diverse demand provides additional trade opportunities for local entrepreneurs.

Schools, training centres, research institutions and extension services can also benefit from this demand for inputs, services and products. When research and development is applied to the innovations being made in home gardens, viable commercial applications often result. Many of the improved varieties of fruit and vegetables popular in markets today were originally selected by observant and skilled home gardeners.

DEVELOPING HOME GARDENS

Decision Making

Choosing what to produce for a family and for sale needs to involve all family members. A home garden employs and feeds almost all family members, therefore women, men and the elderly should be involved together in decisions about a home garden. More produce is consumed by women and children, especially young children, from gardens controlled by women. Choices are best made considering the whole farming system, since garden products tend to complement other farm production.

Selection of what to produce for the family should take into account nutritional needs and family cultural and traditional preferences. Nutritious food may not be consumed if people do not like to eat it. Traditional mixed subsistence gardens can supply diverse nutritional needs and provide many of a family's favourite herbs, spices and flavours. Producing additional food crops spreads risk and may provide a surplus for sale. In all but the most remote areas, there are market opportunities for selected cash crops. Poultry and livestock can provide good cash income, either from fattening for meat or from regular sales of eggs, milk or fibre.

Caution should be exercised in promoting or replacing traditional and indigenous crops. They may serve a historic need that has been superseded by more economically efficient alternatives, such as corrugated roofing steel replacing demand for thatching palms. Local use of traditional plants may not be immediately obvious to an outsider, but it will remain important to household preferences. Tourism potential should not be overlooked. Tourist visits to rural communities are popular; tourists like to observe traditional village industries, handicrafts and specialisations such as cultivation of spices, medicinal plants, flowers and ornamental plants.

Market Choices

Market choices are important; marketing is examined in more detail later. Production for market is of two kinds: sale of produce surplus to family needs, and produce grown specifically for markets. Neighbours in a village and local community are markets that are often ignored, although they are easy to reach. With a more complex marketing chain using transport and other resources, a home gardener may aim to supply larger markets in nearby and more distant population centres. These markets require a more specialised garden system or a collective approach to provide sufficient quality, volume and shipment frequency.

Finding Resources

Extra resources increase the options for improving production, but gardeners must assess what they have access to and what suits their situation. Time-consuming operations can be completed faster and more easily with machinery, which makes it possible to extend a garden into a smallholding. To mechanise operations, choices need to be made about power sources whether hand, pedal or engine power, and how to obtain access to them. Available sources of energy need to be considered, such as solar energy for drying home-processed foods, or biogas captured from manure compost in a home garden to be used for cooking. Wind and water can be harnessed to power mills, water pumps and generators.

Mechanisation options depend on access to equipment, operator training services, spare parts and maintenance services. Access to financial resources such as savings, informal credit or bank loans and technical and commercial information increase potential options to develop home gardens.

Potential Constraints

Government policies can constrain home garden potential. Land-use restrictions at the rural-urban interface or as part of state-owned land lease conditions can limit certain production systems or choices of crops. Health policies can restrict gardeners' access to higher-priced markets, Some countries require vegetables to be washed with a chlorine solution, the waste from which is environmentally damaging and inappropriate for disposal in a home garden. Use of manures and compost manufacture can cause clashes with planning authorities and neighbours. Increased gardening can have negative social implications, from environmental changes brought about by the increased intensity of home gardening to changes in the household.

Control of home gardens is an important incentive, especially control of income from sale of garden produce. In some areas of Africa, men have taken over management and marketing choices as gardens managed by women have become more profitable, which not only decreases cash incentives for women but may also decrease the nutritional value of a garden. Establishing a home garden requires labour and resources, sometimes including capital. In the rural environment of developing countries, many household needs compete for cash and other resources. Home garden systems should be developed that minimise risks. It is essential to ensure that gardens are sustainable. Home gardens that rely on externally-supplied inputs such as seeds or greenhouse polythene may fail if gardeners are not trained to save their own seeds, or if external support for access to input supply markets is withdrawn.

Low Jome-garden Production

Success is more likely with a people-centred, interdisciplinary approach that develops and improves existing technologies. Indigenous production systems should not be disrupted by "introduction of new and more efficient methods". Gardens, like smallholder farming, are part of social systems. It is best to develop home gardens that meet family needs and resources rather than the ideals of specialists in agriculture, health or land-use planning.

Design

Once a good understanding of current gardening is established, households and promoters can design improvements to make the garden supply family nutrition or production requirements. Improvements may take the form of introducing a wider diversity of plants, animals and fish, more effective management practices or opening up a new space for a special type of enterprise such as animal rearing or fish farming. It is essential at this stage to build on indigenous knowledge, especially with regard to native and wild plants—forests and other non-farm areas may seasonally supply greens, spices and mushrooms—pest and disease management and mixed-cropping systems.

Space

Improving traditional garden space usually involves one or more of the following: replacing plants, increasing planting density, introducing new components such as livestock or increasing external inputs, especially water and nutrients. Multiple cropping and multi-layer systems utilise existing light and space; examples are beans or pumpkins together with maize or sorghum, climbing vines on canopy trees or root crops beneath fruit trees. Replacement is possible where traditional gardens contain crops that have outlived demand. In areas where steel roofing has taken over from thatched roofs, for example, it makes sense to replace old thatching palm grown as a cash crop for higher-value crops.

Layout

Improving the layout involves integrating garden features so that efficient and sustainable use is made of structures, land forms, water and other materials. One of the first concerns is security, especially against wandering poultry and livestock. Many successful gardens make maximum use of local materials and resources. Living fences strengthen garden fencing; many, such as the multipurpose tree *Gliricida sepium*, also yield leaves for animal fodder or compost and sticks for firewood. The location of kitchen gardens next to a house provides some fencing needs and improves the efficiency of labour and resources.

Resources

Improved gardens typically make the best use of locally available resources. Layout generates synergy between social and biological elements. Livestock, when controlled or selectively excluded from plant production areas, provide manure for fertiliser; goats will clear areas for planting and poultry will eat weed seedlings and insects among tolerant crops. Flowers and other plants may enhance the social functions of a home garden and play a role as biofilters in reducing pests—marigolds are an example -or act as a living food store—canna, *Canna edulis,* is an example. Low-lying areas and steeply sloping land may be unsuitable for field agriculture, but may be developed into a sustainable garden using small-scale land-forming techniques such as raised beds, terraces and hedgerows. Wastes are a valued resource in many home gardens. Household and post-harvest waste can be minimised by recycling water and nutrients. This is a fundamental concept of permaculture gardens such as the vegetable, aquaculture and animal cooperative (VAC) garden system in Viet Nam.

Tools and equipment

The vast majority of home gardeners worldwide use only hand tools. Typical home gardeners have tools such as hoes or spades for soil preparation, rakes, forks, baskets and barrows for handling materials and sometimes dibbling sticks for planting seed. A range of equipment, depending on resources available, is used in developing countries to assist production and processing operations carried out in home gardens. Examples of more advanced equipment are listed below.

- Polythene-film traps that warm air around tender plants allow vegetables to be grown in the off-season. In Bhutan, gardeners grow pumpkin at elevations of 3 000 m using plastic cloches to prevent snow damage in spring. In tropical areas, shade cloths protect young seedlings from heat stress. Where space is limited, nets and plastic film are used as fences to keep out village poultry and field rats, and they do not compete with plants for light, water and nutrients.
- Small water pumps and lifting devices reduce time and energy in irrigation.
- Motor tillers primarily used in other components of the farming system are sometimes used for soil preparation and transport.
- Much crop processing is done in the home garden. Cleaning or hulling equipment for maize or coffee, for example, can be manual or electric. Small dryers for grain, fruit and vegetables utilise solar energy, firewood or electricity. Milling equipment is a favourite amongst women to replace the daily chore of pounding grain.

Supporting Resources for Home Gardening

All farming systems that produce output for sale need some physical inputs such as initial seeds and regular nutrients to replace those exported. These inputs must be bought and paid for in one way or another. Home gardeners employ a range of sustainable strategies to avoid having to pay, including saving what they have and recycling wastes. To provide vegetable seed for later crops, some plants are left to flower and seed. Such "land race" seed may not provide the vigour or the market benefits of commercial hybrid seed, but the gardener does not have to spend cash or rely on market access. Good gardeners select plants with the best production and market characteristics to be nurtured into strong seed, for which there may be a good local market if any surplus seed is produced.

The largest and most costly inputs in intensive agriculture are often fertilisers and feeds. Nitrogen-fixing plants such as beans and multipurpose trees are effective on-site resources of nutrients as a green manure, or as a component of animal feed. Utilising unwanted resources was described earlier as an important design feature in the layout of improved gardens. Kitchen gardens can survive on waste water and organic household and farm wastes, which ideally are composted to provide nutrients for the garden and often for field crops. Animal manure is a valuable resource in developed and developing countries. Environmental impact increases as garden enterprise levels become more intense. Waste management is an important concern in densely populated areas. Where local environmental laws and government practices allow, gardeners can make a positive contribution through composting, recycling and animal feeding, providing an end use for wastes.

Energy

The range and intensity of income-generating activities possible for individuals or groups of home garden enterprises are dependent on energy, as well as nutrients and other inputs. Resources may already be available in communities, such as biomass from crop production or processing residues, or firewood grown on-site in a living garden fence. Renewable energy technologies allow garden enterprises a degree of independence; they are often the most appropriate technologies for smallholders. Water pumps, for example, can be powered by gravity using a hydraulic ram or by wind, people, livestock or petroleum fuels.

Transport

Access to appropriate transport is a critical resource for viable home garden enterprises for receiving inputs and taking produce to be sold in distant markets. Where roads or tracks are rough or for foot traffic only, it may not be possible to sell fresh produce in distant markets; non-perishable items or local processing is favoured by successful entrepreneurs. Where roads are reasonable, baskets of produce can be taken to market by local bus, in hand carts or horse carts, on motorcycles or by local taxi. Alternatively, gardeners take advantage of visiting traders coming to buy or sell field crops or cash crops. Once a village is known for its good home garden produce, traders often follow.

Support programmes and networks

People with useful expertise and experience are to be found in most communities. They can often be utilised to meet training and skills requirements and other resources, either as individuals or as members of institutions such as farmer groups, women's groups or local government. Experience of home-garden development projects shows that selection and training of "master gardeners" as community garden promoters is the best way to ensure viable gardens and retain indigenous knowledge in planning home garden improvements. In some countries, there is public-sector support for small or medium enterprise development, which covers processing activities, trading and services. This kind of support facilitates commercialisation by providing entrepreneurs with training in business planning, assistance in establishing group trading or processing contracts and support for accessing financial credit.

Formal training opportunities for improved home gardening are typically limited by tight public sector budgets, but experienced local people, especially retired people, are important informal training resources. Technical information is available from agriculture departments and gardeners' groups. In ethnically diverse countries, pictures and graphics are used in technical posters and leaflets to overcome literacy constraints.

There is good potential for taking advantage of this by establishing groups of home gardeners, cooperatives and similar organisations to exploit resources of information, technology and markets. Many countries already have specialised support networks such as gardeners' clubs or associations, fruit growers' or farmers' federations or unions that can provide training, supplies and access to model gardens. Community-development organisations and groups sponsored by communities or local governments, such as women's groups and social forestry groups, often provide facilitation.

Marketing

Marketing decisions

It is a household decision whether to consume garden produce in the home or sell it into the community—but it is always the market that decides what to buy, when to buy it and how much it will pay for it. Produce grown for family subsistence may not be in demand by other families in a time of plenty. On the other hand, well stored products sold during a shortage period may earn good prices. A buyer might decide that what is offered for sale is of better quality than anything else in the market and buy, even at prices higher than those for lower-quality produce. Markets thus determine choices of crop or livestock that can be produced for sale.

In choosing what market to supply, gardeners have to decide if they can meet market needs. Apart from the type of product and its quality, they need to consider how to get their product into the market and when to supply it; appropriate transport and supply processes are essential. More buyers may choose a product that is promoted so that they are familiar with it.

Market information

Decisions on crop and animal types, production schedules, produce quality and organisation of distribution are improved with better information. Market information is a significant factor in improving home garden enterprises. Gardeners use four main kinds of information to help them fine tune production and post-harvest handling practices.

- The size of a market helps gardeners estimate how much they should produce and sell; this information includes the quantity or weight to be sold, the number of buyers and how much they buy.
- Information about the duration of demand for a product, for example, staple foods consumed year round or flowers for festive or holy days, is used to fine tune production scheduling, product storage and distribution systems.
- Information about the popular characteristics of a product that buyers prefer including size, shape, weight, colour, packaging and overall quality, enables gardeners to enjoy premium prices and loyal, regular buyers. Market information often includes the range of prices for different grades of product quality.
- The location of the market requires gardeners to develop appropriate packing, handling and transport procedures.

Finding out market information helps to identify opportunities for sales of fresh produce and outlets for sales of value-added or finished products. Cash crops such as coffee or vanilla sell for better prices if they are properly processed and dried, which may be done by individual households or done on contract by a villager who has the appropriate equipment. Home gardeners learn how to exploit small-scale markets in their communities. Once farmers are familiar with these and generating some income, their gardens may evolve into smallholder enterprises supplying larger and more distant markets.

Serving customers

One of the most significant aspects of successful marketing is the importance of the client. Successful small-scale enterprises meet market needs. In some countries, especially those in transition from a command economy to a free market, the successful people are those who find out what their customers want. Marketing new kinds of produce may require adaptation and promotion to expand local preferences; some customers may want to ensure that they can get a regular supply. Generally, the higher the incomes of the buyers, the higher their demand for quality. To gain a share of more sophisticated markets, gardeners must supply better-quality produce, which will often come from high-quality varieties and improved livestock breeds.

Maintaining supply

To maintain supplies of produce that suits the customers, home garden enterprises need experience and knowledge of the chain of events from production and processing to distribution and sales. Vegetable growers ensure they can provide a regular supply by leaving a week or more between plantings. Processing markets for cash crops have specific requirements: some, such as fruit processors, need to spread supply over long periods to maintain factory throughput. High-quality coffee processors need to have coffee cherries delivered to them on the day the cherries are harvested.

Packaging and means of transport differ according to how, when and where the produce is wanted. Nurseries produce plants either as bare-root plants or in pots or tubes. Bare-root plants are generally trees and shrubs for local planting in the cool season. Bare-root plants may be cheaper and more appropriate in certain situations such as farm woodlots, but potted plants are easier to transport and may arrive in better condition. This is especially important for nursery gardens serving intensive field agriculture, where farmers want bulk crop seedlings of uniform standard to give them a seasonal head start.

Successful small-scale entrepreneurs need to take care that their input supply system is reliable and efficient. Animal breeders need to ensure access to feed, veterinary services and products. Seed and other propagation material may be available through traders, shops or by seasonal collection from other farmers. Mushroom growers, for example, need stable supplies of straw, sawdust or other growth substrates.

REFERENCES

Ball, Jeff. 1988. *Rodales's garden problem solver: vegetables, fruits, and herbs.*Emmaus, PA: Rodale Press.

Berbiglia, Julie. 1996. *The Lazy Gardener'sGuide to Organic Gardening.* Scarritt-Bennett.

Bradley, Fern marshall and Barbara W.Ells, editors. 1992. *Rodale's all-new encyclopidia of organic gardening: the indispensable resource for every gardener.* Emmaus, PA: Rodale Press,

Crockett, James Underwood. 1972. *Vegetables and Fruits.* New York: Time Life Books.

Denckla, Tanya. 1994. *The Organic Gardener's Home Reference: a plant-by-plant guide to growing fresh, healthy food.* Pownal, VT: Storey Communications.

Doty, Walter L. 1990. *All about vegetables.* San Ramon, CA: Ortho Books.

Hart, Rhonda Massingham. 1991. *Bugs, slugs & other thugs: controlling garden pests organically.* Pownal, VT:Storey Communications.

Lawton, Barbara P.and George F.Van Patten. 1993. *Organic gardener'sbasics.* Van Patten Publishing.

Lima, Patrick. *The natural food garden: growing vegetables and fruits chemical-free.* 1992. Rocklin, CA: Prima Pub.

9

Horticultural Practices

Horticultural practice naturally range under the three heads of flowers, fruits and vegetables. There are, however, certain general aspects of the subject which will be more conveniently noticed apart, since they apply alike to each department. Therefore first treat of these under four headings: formation and preparation of the garden, garden structures and edifices, garden materials and appliances, and garden operations.

PREPARATION OF HORTICULTURE SITE

The site chosen for the mansion will more or less determine that of the garden, the pleasure grounds and flower garden being placed so as to surround or lie contiguous to it, while the fruit and vegetable gardens, either together or separate, should be placed on one side or in the rear, according to fitness as regards the nature of the soil and subsoil, the slope of the surface or the general features of the park scenery. In the case of villa gardens there is usually little choice: the land to be occupied is cut up into plots, usually rectangular, and of greater or less breadth, and in laying out these plots there is generally a smaller space left in the front of the villa residence and a larger one behind, the front plot being usually devoted to approaches, shrubbery and plantations, flower beds being added if space permits, while the back or more private plot has a piece of lawn grass with flower beds next the house, and a space for vegetables and fruit trees at the far end, this latter being shut off from the lawn by an intervening screen of evergreens or other plants. Between these two classes of gardens there are many gradations, but our remarks will chiefly apply to those of larger extent.

The almost universal practice is to have the fruit and vegetable gardens combined; and the flower garden may sometimes be conveniently placed in juxtaposition with them. When the fruit and vegetable gardens are combined, the smaller and choicer fruit trees only should be admitted, such larger-growing hardy fruits as apples, pears, plums, cherries, being relegated to the orchard. Ground possessing a gentle inclination towards the south is desirable for a garden.

On such a slope effectual draining is easily accomplished, and the greatest possible benefit is derived from the sun's rays. It is well also to have an open exposure towards the east and west, so that the garden may enjoy the full benefit of the morning and evening sun, especially the latter; but shelter is desirable on the north and north-east, or in any direction in which the particular locality may happen to be exposed. In some places the south-western gales are so severe that a belt of trees is useful as a break wind and shelter.

Soil and Subsoil

A hazel-coloured loam, moderately light in texture, is well adapted for most garden crops, whether of fruits or vegetables, especially a good warm deep loam resting upon chalk; and if such a soil occurs naturally in the selected site, but little will be required in the way of preparation. If the soil is not moderately good and of fair depth, it is not so favourable for gardening purposes. Wherever the soil is not quite suitable, but is capable of being made so, it is best to remedy the defect at the outset by trenching it all over to a depth of 2 or 3 ft., incorporating plenty of manure with it. A heavy soil, although at first requiring more labour, generally gives far better results when worked than a light soil.

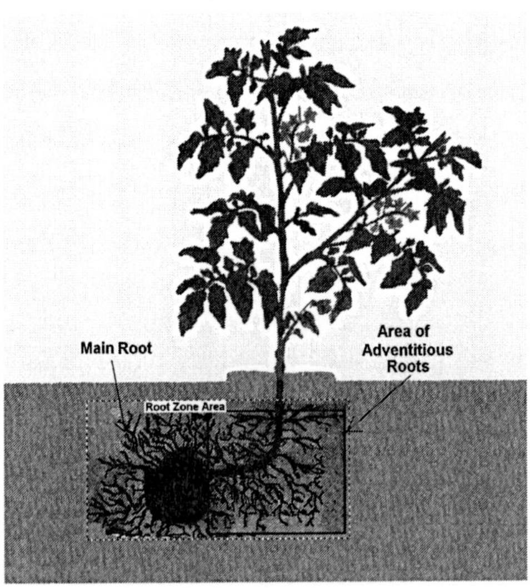

Figure 1. A graden tomato plant in good soil

The latter is not sufficiently retentive of moisture and gets too hot in summer and requires large quantities of organic manures to keep it in good condition. It is advantageous to possess a variety of soils; and if the garden be on a slope it will often be practicable to render the upper part light and dry, while the lower remains of a heavier and damper nature. Natural soils consist of substances derived from the decomposition of various kinds of rocks, the bulk consisting of clay, silica and lime, in various proportions. As regards preparation, draining is of course of the utmost importance.

The ground should also be trenched to the depth of 3 ft. at least, and the deeper the better so as to bring up the subsoil—whether it be clay, sand, gravel, marl. Jor exposure to the weather and thus convert it from a sterile mass into a living soil teeming with bacteria. In this operation all stones larger than a man's fist must be taken out, and all roots of trees and of perennial weeds carefully cleared away. When the whole ground has been thus treated, a moderate liming will, in general, be useful, especially on heavy clay soils. After this, supposing the work to have occupied most of the summer, the whole may be laid up in ridges, to expose as great a surface as possible to the action of the winter's frost.

Argillaceous or clay soils are those which contain a large percentage (45-50) of clay, and a small percentage (5 or less) of lime. These are unfitted for garden purposes until improved by draining, liming, trenching and the addition of porous materials, such as ashes, burnt ballast or sand, but when thoroughly improved they are very fertile and less liable to become exhausted than most other soils. Loamy soils contain a considerable quantity (30-45%) of clay, and smaller quantities of lime, humus and sand. Such soils properly drained and prepared are very suitable for orchards, and when the proportion of clay is smaller (20-30%) they form excellent garden soils, in which the better sort of fruit trees luxuriate.

Manly soils are those which contain a considerable percentage (10-20) of lime, and are called clay marls, loamy marls and sandy marls, according as these several ingredients preponderate. The clay marls are, like clay soils, too stiff for garden purposes until well worked and heavily manured; but loamy marls are fertile and well suited to fruit trees, and sandy marls are adapted for producing early crops.

Calcareous soils, which may also be heavy, intermediate or light, are those which contain more than 20% of lime, their fertility depending on the proportions of clay and sand which enter into their composition; they are generally cold and wet. Vegetable soils or moulds, or humus soils, contain a considerable percentage (more than 5) of humus, and embrace both the rich productive garden moulds and those known as peaty soils. The nature of the subsoil is of scarcely less importance than that of the surface soil.

Many gardeners are still afraid to disturb an unsuitable subsoil, but experienced growers have proved that by bringing it up to the surface and placing plenty of manure in the bottoms of the various trenches, the very best results are attained in the course of a season or so. An uneven subsoil, especially if retentive, is most undesirable, as water is apt to collect in the hollows, and thus affect the upper soil. The remedy is to make the plane of its surface agree with that of the ground. When there is a hard pan this should be broken up with the spade or the fork, and have plenty of manure mixed with it. When there is an injurious preponderance of metallic oxides or other deleterious substances, the roots of trees would be affected by them, and they must therefore be removed. When the subsoil is too compact to be pervious to water, effectual drainage must be resorted to; when it is very loose, so that it drains away the fertile ingredients of the soil as well as those which are artificially supplied, the compactness of the stratum should be increased by the addition of clay, marl or loam. The best of all subsoils is a dry bed of clay overlying sandstone.

Plan

In laying out the garden, the plan should be prepared in minute detail before commencing operations. The form of the kitchen and fruit garden should be square or oblong, rather than curvilinear, since the working and cropping of the ground can thus be more easily carried out. The whole should be compactly arranged, so as to facilitate working, and to afford convenient access for the carting of the heavy materials. This access is especially desirable as regards the store-yards and framing ground, where fermenting manures and tree leaves for making up hot beds, coals or wood for fuel and ingredients for composts, together with flower-pots and the many necessaries of garden culture, have to be accommodated.

Figure 2. Proper location of a home graden

In the case of villas or picturesque residences, gardens of irregular form may be permitted; when adapted to the conditions of the locality, they associate better with surrounding objects, but in such gardens wall space is usually limited. The distribution of paths must be governed by circumstances. Generally speaking, the main paths for cartage should be 8 ft. wide, made up of 9 in. hard core covered by 4 in. of gravel or ash, with a gentle rise to centre to throw off surface water. The smaller paths, not intended for cartage, should be 4 ft. to 6 ft. wide, according to circumstances, made up of 6 in. hard core and 3 in. of gravel or ash, and should be slightly raised at centre. A considerable portion of the north wall is usually covered in front with the glazed structures called forcing-houses, and to these the houses for ornamental plants are sometimes attached; but a more appropriate site for the latter is the flower garden, when that forms a separate department. It is well, however, that everything connected with the forcing of fruits or flowers should be concentrated in one place.

The frame ground, including melon and pine pits, should occupy some well-sheltered spot in the slips, or on one side of the garden, and adjoining to this may be found a suitable site for the compost ground, in which the various kinds of soils are kept in store, and in which also composts may be prepared. As walls afford valuable space for the growth of the choicer kinds of hardy fruits, the direction in which they are built is of considerable importance.

Figure 3. Crops in a well-structured garden

In the warmer parts of the country the wall on the north side of the garden should be so placed as to face the sun at about an hour before noon, or a little to the east of south; in less favoured localities it should be made to face direct south, and in still more unfavourable districts it should face the sun an hour after noon, or a little west of south. The east and west walls should run parallel to each other, and at right angles to that on the north side, in all the most favoured localities; but in colder or later ones, though parallel, they should be so far removed from a right angle as to get the sun by eleven o'clock. On the whole, the form of a parallelogram with its longest sides in the proportion of about five to three of the shorter, and running east and west, may be considered the best form, since it affords a greater extent of south wall than any other.

Shelter

A screen of some kind to temper the fury of the blast is absolutely necessary. If the situation is not naturally well sheltered, the defect may be remedied by masses of forest trees disposed at a considerable distance so as not to shade the walls or fruit trees. They should not be nearer than, say, 50 yds., and may vary from that to ioo or 150 yds. distance according to circumstances, regard being had especially to peculiarities occasioned by the configuration of the country, as for instance to aerial currents from adjacent eminences. Care should be taken, however, not to hem in the garden by crowded plantations, shelter from the prevailing strong winds being all that is required, while the more open it is in other directions the better.

Figure 4. Crops under shelter

The trees employed for screens should include both those of deciduous and of evergreen habit, and should suit the peculiarities of local soil and climate. Of deciduous trees the sycamore, wych-elm, horse-chestnut, beech, lime, plane and poplar may be used,—the abele or white poplar, Populus alba, being one of the most rapidgrowing of all trees, and, like other poplars, well suited for nursing other choicer subjects; while of evergreens, the holm oak, holly, laurel, and such conifers as the Scotch, Weymouth and Austrian pines, with spruce and silver firs and yews, are suitable. The conifers make the most effective screens.

Figure 5. Kale growing in the vegetable garden.

Extensive gardens in exposed situations are often divided into compartments by hedges, so disposed as to break the force of high winds. Where these are required to bc narrow as well as lofty, holly, yew or beech is to be preferred; but, if there is sufficient space, the beautiful laurel and the bay may be employed where they will thrive. Smaller hedges may be formed of evergreen privet or of tree-box. These subordinate divisions furnish, not only shelter but also shade, which, at certain seasons, is peculiarly valuable. Belts of shrubbery may be placed round the slips outside the walls; and these may in many cases, or in certain parts, be of sufficient breadth to furnish pleasant retired promenades, at the same time that they serve to mask the formality of the walled gardens, and are made to harmonise with the picturesque scenery of the pleasure ground.

Water Supply

Although water is one of the most important elements in plant life, we do not find one garden in twenty where even ordinary precautions have been taken to secure a competent supply. Rain-water is the best, next to that river or pond water, and last of all that from springs; but a chemical analysis should be made of the last before introducing it, as some spring waters contain mineral ingredients injurious to vegetation.

Figuer 6. Watering crops with pipe system

Iron pipes are the best conductors; they should lead to a capacious open reservoir placed outside the garden, and at the highest convenient level, in order to secure sufficient pressure for effective distribution, and so that the wall trees also may be effectually washed. Stand-pipes should be placed at intervals beside the walks and in other convenient places, from which water may at all times be drawn; and to which a garden hose can be attached, so as to permit of the whole garden being readily watered. The mains should be placed under the walks for safety, and also that they may be easily reached when repairs are required. Pipes should also be laid having a connexion with all the various greenhouses and forcing-houses, each of which should be provided with a cistern for aerating the daily supplies. In fact, every part of the garden, including the working sheds and offices, should have water supplied without stint.

Fence

Gardens of large extent should be encircled by an outer boundary, which is often formed by a sunk wall or ha-ha surrounded by an invisible wire fence to exclude ground game, or consists of a hedge with low wire fence on its inner side. Occasionally this sunk wall is placed on the exterior of the screen plantations, and walks lead through the trees, so that views are obtained of the adjacent country.

Although the interior garden receives its form from the walls, the ring fence and plantations may be adapted to the shape and surface of the ground. In smaller country gardens the enclosure or outer fence is often a hedge, and there is possibly no space enclosed by walls, but some divisional wall having a suitable aspect is utilised for the growth of peaches, apricots, and the hedge merely separates the garden from a paddock used for grazing. The still smaller gardens of villas are generally bounded by a wall or wood fence, the inner side of which is appropriated to fruit trees. For the latter walls are much more convenient and suitable than a boarded fence, but in general these are too low to be of much value as aids to cultivation, and they are best covered with bush fruits or with ornamental plants of limited growth.

Walks

The best material for the construction of garden walks is good binding gravel. The ground should be excavated to the depth of a foot or more—the bottom being made firm and slightly concave, so that it may slope to the centre, where a drain should be introduced; or the bottom may be made convex and the water allowed to drain away at the sides. The bottom g in. should be filled in compactly with hard, coarse materials, such as stones, brickbats, clinkers, burned clay, on which should be laid 2 or 3 in. of coarse gravel, and then r or 2 in. of firm binding gravel on the surface. The surface of the walks should be kept well rolled, for nothing contributes more to their elegance and durability.

All the principal lines of walk should be broad enough to allow at least three persons to walk abreast; the others may be narrower, but a multitude of narrow walks has a puny effect. Much of the neatness of walks depends upon the material of which they are made. Gravel from an inland pit is to be preferred; though occasionally very excellent varieties are found upon the sea-coast.

Gravel walks must be kept free from weeds, either by hand weeding, or by the use of one of the many weed killers now on the market. In some parts of the country the available material does not bind to form a close, even surface, and such walks are kept clean by hoeing. Grass walks were common in English gardens during the prevalence of the Dutch taste, but, owing to the frequent humidity of the climate, they have in a great measure been discarded.

Grass walks are made in the same way as grass lawns. When the space to be thus occupied is prepared, a thin layer of sand or poor earth is laid upon the surface and over this a similar layer of good soil. This arrangement is adopted in order to prevent excessive luxuriance in the grass. In many modern gardens pathways made of old paving stones lead from the house to different parts. They give an old-fashioned and restful appearance to a garden, and in the interstices charming little plants like thyme, Ionopsidium acaule, are allowed to grow.

Edgings

Walks are separated from the adjoining beds and borders in a variety of ways. If a living edging is adopted, by far the best is afforded by the dwarf box planted closely in line. It is of extremely neat growth, and when annually clipped will remain in good order for many years. Very good edgings, but of a less durable character, are formed by thrift. (*Armeria vulgaris*), double daisy (*Bellis perennis*), gentianella. (*Gentiana acaulis*) and London pride (*Saxifrage umbrosa*), *Cerastium tomentosum, Stachys lavata* and the beautiful evergreen *Veronica rupestris* with sheets of bright blue flowers, close to the ground, or by some of the finer grasses very carefully selected, such as the sheep's fescue (*Festuca ovina*) or its glaucous-leaved variety. Indeed, any low-growing herbaceous plant, susceptible of minute division, is suitable for an edging.

Amongst shrubby plants suitable for edgings are the evergreen candytuft (*Iberis sempervirens*), *Euonymus radicans variegata*, ivy, and *Euonymus microphyllus*—a charming little evergreen with small serrated leaves. Edgings may also be formed of narrow slips of sandstone flag, slate, tiles or bricks.

One advantage of using edgings of this kind, especially in kitchen gardens, is that they do not harbour slugs and similar vermin, which all live edgings do, and often to a serious extent, if they are left to grow large. In shrubberies and large flower-plots, verges of grass-turf, from r to 3 ft. in breadth, according to the size of the border and width of the walk, make a very handsome edging, but they should not be allowed to rise more than an inch and a half above the gravel, the grass being kept short by repeated mowings, and the edges kept trim and well-defined by frequently clipping with shears and cutting once or twice a year with an edging iron.

GARDEN STRUCTURES

Walls

The position to be given to the garden walls has been already referred to. The shelter afforded by a wall, and the increased temperature secured by its presence, are indispensable in the climate of Great Britain, for the production of all the finer kinds of outdoor fruits; and hence the inner side of a north wall, having a southern aspect, is appropriated to the more tender kinds. It is, indeed, estimated that such positions enjoy an increased temperature equal to 7° of latitude—that is to say, the mean temperature within a few inches of the wall is equal to the mean temperature of the open plain 7° farther south. The eastern and western aspects are set apart for fruits of a somewhat hardier character. Where the inclination of the ground is considerable, and the presence of high walls would be objectionable, the latter may be replaced by sunk walls. These should not rise more than 3 ft. above the level of the ground behind them.

As dryness is favourable to an increase of heat, such walls should be either built hollow or packed behind to the thickness of 3 or 4 ft. with rubble stones, flints, brickbats or similar material, thoroughly drained at bottom. For mere purposes of shelter a height of 6 or 7 ft. will generally be sufficient for the walls of a garden, but for the training of fruit trees it is found that an average height of 12 ft. is more suitable. In gardens of large size the northern or principal wall may be 14 ft., and the side walls 12 ft. in height; while smaller areas of an acre or so should have the principal walls 12 and the side walls io ft. in height. As brick is more easily built hollow than stone, it is to be preferred for garden walls. A 14-in. hollow wall will take in its construction 12,800 bricks, while a solid 9-in. one, with piers, will take 11,000; but the hollow wall, while thus only a little more costly, will be greatly superior, being drier and warmer, as well as more substantial.

Bricks cannot be too well burnt for garden walls; the harder they are the less moisture will they absorb. Many excellent walls are built of stone. The best is dark-coloured whinstone, because it absorbs very little moisture, or in Scotland Caithness pavement 4 in. thick. The stones can be cut to any required length, and built in regular courses. Stone walls should always be built with thin courses for convenience of training over their surface. Concrete walls, properly coped and provided with a trellis, may in some places be cheapest, and they are very durable. Common rubble walls are the worst of all. The coping of garden walls is important, both for the preservation of the walls and for throwing the rain-water off their surfaces. It should not project less than from 2 to 22 in., but in wet districts may be extended to 6 in.

Stone copings are best, but they are costly, and Portland cement is sometimes substituted. Temporary copings of wood, which may be fixed by means of permanent iron brackets just below the stone coping, are extremely useful in spring for the protection of the blossoms of fruit trees. They should be 9 in. or 1 ft. wide, and should be put on during spring before the blossom buds begin to expand; they should have attached to them scrim cloth, which admits light pretty freely, yet is sufficient to ward off ordinary frosts; this canvas is to be let down towards evening and drawn up again in the morning.

These copings should be removed when they are of no further utility as protectors, so that the foliage may have the full benefit of rain and dew. Any contrivance that serves to interrupt radiation, though it may not keep the temperature much above freezing, will be found sufficient. Standard fruit trees must be left to take their chance; and, indeed from the lateness of their flowering, they are generally more injured by blight, and by drenching rains, which wash away the pollen of the flowers, than by the direct effects of cold.

Espalier Rails

Subsidiary to walls as a means of training fruit trees, espalier rails were formerly much employed, and are still used in many gardens. In their simplest form, they are merely a row of slender stakes of larch or other wood driven into the ground, and connected by a slight rod or fillet at top. The use of iron rails has now been almost wholly discontinued on account of metallic substances acting as powerful conductors of both heat and cold in equal extremes. Standards from which galvanised wire is tightly strained from one end to the other are preferable and very convenient. Trees trained to them are easily got at for all cultural operations, space is saved, and the fruit, while freely exposed to sun and air, is tolerably secure against wind. They form, moreover, neat enclosures for the vegetable quarters, and, provided excess of growth from the centre is successfully grappled with, they are productive in soils and situations which are suitable.

Plant Houses

These include all those structures which are more intimately associated with the growth of ornamental plants and flowers, and comprise conservatory, plant stove, greenhouse and the subsidiary pits and frames. They should be so erected as to present the smallest extent of opaque surface consistent with stability. With this object in view, the early improvers of hot-house architecture substituted metal for wood in the construction of the roofs, and for the most part dispensed with back walls; but the conducting power of the metal caused a great irregularity of temperature, which it was found difficult to control; and, notwithstanding the elegance of metallic houses, this circumstance, together with their greater cost, has induced most recent authorities to give the preference to wood.

The combination of the two, however, shows clearly that, without much variation of heat or loss of light, any extent of space may be covered, and houses of any altitude constructed. Plant houses must be as far as possible impervious to wet and cold air from the exterior, provision at the same time being made for ventilation, while the escape of warm air from the interior must also be under control.

The most important part of the enclosing material is necessarily glass. But as the rays of light, even in passing through transparent glass, lose much of their energy, which is further weakened in proportion to the distance it has to travel, the nearer the plant can be placed to the glass the more perfectly will its functions be performed; hence the importance of constructing the roofs at such an angle as will admit the most light, especially sunlight, at the time it is most required.

Plants in glass houses require for their fullest development more solar light probably than even our best hot-houses transmit—certainly much more than is transmitted through the roofs of houses as generally constructed. Plant houses constructed of the best Baltic pine timber are very durable, but the whole of the parts should be kept as light as possible. In many houses, especially those where ornament is of no consequence, the rafters are now omitted, or only used at wide intervals, somewhat stouter sash-bars being adopted, and stout panes of glass 12 to 18 in. wide, made use of. Such houses are very light; being also very close, they require careful ventilation.

The glass roof is commonly designed so as to form a uniform plane or slope from back to front in lean-to houses, and from centre to sides in span-roofed houses. To secure the greatest possible influx of light, some horticulturists recommend curvilinear roofs; but the superiority of these is largely due to the absence of rafters, which may also be dispensed with in plain roofs. They are very expensive to build and maintain. Span and ridge-and-furrow roofs, the forms now mostly preferred, are exceedingly well adapted for the admission of light, especially when they are glazed to within a few inches 2 of the ground. They can be made, too, to cover in any extent of area without sustaining walls. Indeed, it has been proposed to support such roofs to a great extent upon suspension principles, the internal columns of support being utilised for conducting the rain-water off the roof to underground drains or reservoirs.

The lean-to is the least desirable form, since it scarcely admits of elegance of design, but it is necessarily adopted in many cases. In glazing, the greater the surface of glass, and the less space occupied by rafters and astragals as well as overlaps, the greater the admission of light. Some prefer that the sash-bars should be grooved instead of rebated, and this plan exposes less putty to the action of the weather.

The simple bedding of the glass, without the use of over putty, seems to be widely approved; but the glass may be fixed in a variety of other ways, some of which are. patented. The Conservatory is often built in connexion with the mansion, so as to be entered from the drawing-room or boudoir. But when so situated it is apt to suffer from the shade of the building, and is objectionable on account of admitting damp to the drawing-room. Where circumstances will admit, it is better to place it at some distance from the house, and to form a connexion by means of a glass corridor. In order that the conservatory may be kept gay with flowers, there should be a subsidiary structure to receive the plants as they go out of bloom.

The conservatory may also with great propriety be placed in the flower garden, where it may occupy an elevated terrace, and form the termination of one of the more important walks. Great variety of design is admissible in the conservatory, but it ought always to be adapted to the style of the mansion of which it is a prominent appendage.

Some very pleasing examples are to be met with which have the form of a parallelogram with a lightly rounded roof; others of appropriate character are square or nearly so, with a ridge-and-furrow roof. Whatever the form, there must be light in abundance; and the shade both of buildings and of trees must be avoided. A southern aspect, or one varying to south-east or south-west, is preferable; if these aspects cannot be secured, the plants selected must be adapted to the position. The central part of the house may be devoted to permanent plants; the side stages and open spaces in the permanent beds should be reserved for the temporary plants.

Fruit Houses

The principal of these are the vinery, peach house, cucumber and melon house and orchard house. These, or a portion of them, especially the vineries and peacheries, are frequently brought together into a range along the principal interior or south wall of the garden, where they are well exposed to sun and light, an ornamental plant house being some times introduced into the centre of the range in order to give effect to the outline of the buildings. When thus associated, the houses are usually of the lean-to class, which have the advantage of being more easily warmed and kept warm than buildings having glass on both sides, a matter of great importance for forcing purposes.

The Vinery is a house devoted to the culture of the grape-vine, which is by far the most important exotic fruit cultivated in English gardens. When forming part of a range a vinery would in most cases be a lean-to structure, with a sharp pitch (45°-50°) if intended for early forcing, and a flatter roof (40°) with longer rafters if designed for the main and late crops.

(1) The lean-to is the simplest form, often erected against some existing wall, and the best for early forcing, being warmer on account of the shelter afforded by the back wall. In this house the principal part of the roof is a fixture, ventilation being provided for by small lifting sashes against the back wall, and by the upright front sashes being hung on a pivot so as to swing outwards on the lower side. The necessary heat is provided by four 4-in. hot-water pipes, which would perhaps be best placed if all laid side by side, while the vines are planted in front and trained upwards under the roof. A second set of vines may be planted against the back wall, and will thrive there until the shade of the roof becomes too dense.

(2) The hip-roofed or threequarter span is a combination of the lean-to s and the span-roofed, uniting to a great degree the ad vantages of both, being warmer than the span and lighter than the lean-to. The heating and ventilating arrangements are much the same as in the lean-to, only the top sashes which open are on the back slope, and therefore do not interfere so much with the vines on the front slope. In both this and the lean-to the aspect should be as nearly due south as possible. Houses of this form are excellent for general purposes, and they are well adapted both for muscats, which require a high temperature, and for late-keeping grapes.

(3) The span-roofed, the most elegant and ornamental form, is especially adapted for isolated positions; indeed, no other form affords so much roof space for the development of the vines. The amount of light admitted being very great, these houses answer well for general purposes and for the main crop.

The large amount of glass or cooling surface, however, makes it more difficult to keep up a high and regular temperature in them, and from this cause they are not so well adapted for very early or very late crops. They are best, nevertheless, when grapes and ornamental plants are grown in the same house, except, indeed, in very wet and cold districts, where, in consequence of its greater warmth, the lean-to is to be preferred.

The *Peach House* is a structure in which the ripening of the fruit is accelerated by the judicious employment of artificial heat. For early forcing, as in vineries, the lean-to form is to be preferred, and the house may have a tolerably sharp pitch. A width of 7 or 8 ft., with the glass slope continued down to within a foot or two of the ground, and without any upright front sashes, will be suitable for such a house, which may also be conveniently divided into compartments of from 30 to 50 ft. in length according to the extent of the building, small houses being preferable to larger ones.

As a very high temperature is not required, two or three pipes running the whole length of the house will suffice. The front wall should be built in anticipation of those from the open—walls since a high temperature is not required. A low span, with dwarf side walls, and a lantern ventilator along the ridge, the height in the centre being 9 ft., would be very well adapted for the purpose. The trees should be planted inside and trained up towards the ridge on a trellis about a foot from the glass, the walls being arched to permit the egress of the roots. A trellis path should run along the centre, and movable pieces of trellis should be provided to prevent trampling on the soil while dressing and tying in the young wood.

Pits and Frames

These are used both for the summer growth and winter protection of various kinds of ornamental plants, for the growth of such fruits as cucumbers, melons and strawberries, and for the forcing of vegetables. When heat is required, it is sometimes supplied by means of fermenting dung, or dung and leaves, or tanner's bark, but it is much more economically provided by hot-water pipes. Pits of many different forms have been designed, but it may be sufficient here to describe one or two which can be recommended for general purposes. Span-roof garden frame may under some circumstances be useful as a substitute for the three-light frame. It is adapted for storing plants in winter, for nursing small plants in summer and for the culture of melons and other crops requiring glass shelter.

Frames

Frames should be made of the best red deal, I 4 in. thick. A convenient size is 6 ft. wide, 24 in. high at the back and 15 in front; and they are usually 12 ft. long, which makes three lights and sashes, though they can be made with two lights or one light for particular purposes. Indeed, a one-light frame is often found very convenient for many purposes.

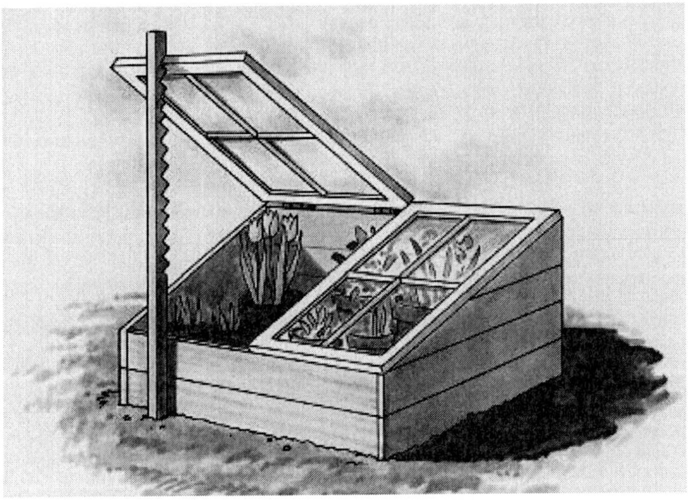

Figure 7. Simple frame for vegetable crops

The lights should be 2 in. thick, and glazed with 21 oz. sheet glass, in broad panes four or five to the breadth of a light, and of a length which will work in conveniently and economically, very long panes being undesirable and with the means of giving sufficient ventilation to keep the air sweet. It should also be sufficiently commodious to permit of the fruit being arranged in single layers on the shelves or trays.

A type of building which is becoming increasingly popular for this purpose, and which is in many respects superior to the older, and often more expensive structures, is built of wood, with or without brick foundations, and is thickly thatched with reeds or other non-conducting material externally—on walls and roof—while the interior is matchboarded. Ventilation is afforded at the ends, usually by tilting laths, operated by a cord. Two doors are provided at one end—an inner, and an outer—the inner being glazed at the top to admit light. They are generally span roofed, about 6 ft. high at the eaves, and 8 or io ft. high at the ridge, according to width.

The length and breadth of these stores should be governed by the amount and character of the storage accommodation to be provided. If intended for storage only, a width of 9 ft. 6 in. would suffice, but if intended to combine display with storage, the internal diameter should be about 13 ft. In the former type, the walls are fitted with four rows of shelves, about 3 ft. wide, and about i ft. 6 in. apart.

The shelves are of deal strips, 2 or 3 in. wide, laid about i in. apart for ventilation. These are being superseded, however, by sliding-out trays of convenient lengths and about 9 in. deep, working on fixed framework. By this means the storage accommodation is nearly doubled and the fruit is more easily manipulated. The central gangway is about 3 ft. 6 in. wide. In the latter a central exhibition bench about 3 ft. wide and of convenient height is provided. Gangways 22 ft. wide flank this, while the shelves or drawers with which the walls are fitted are about 21 ft. wide.

Care of the Fruit Room

This consists mainly in the storing only of such fruits as are dry and in proper condition; in judicious ventilation, especially in the presence of large quantities of newlygathered fruit; in the prompt removal of all decaying fruit; and in the exclusion of vermin. It is also advisable to wash all woodwork and gangways annually with a weak solution of formalin, or other inodorous germicide.

Heating Apparatus

Plant houses were formerly heated in a variety of ways—by fermenting organic matter, such as dung, by smoke flues, by steam and by hot water circulating in iron pipes. The last-named method has proved so satisfactory in practice that it is now in general use for all ordinary purposes. The water is heated by a furnace, and is conveyed from the boiler into the houses by a main or " flow " pipe, connected by means of syphon branches with as many pipes as it is intended to serve. When cooled it is returned to the boiler by another main or " return " pipe. Heat is regulated in the structures by means of valves on the various branch pipes.

The flow pipe is attached to the boiler at its highest point, to take the heated water as it ascends. The return pipe is connected with the boiler at or near its lowest point. The highest points of the pipes are fitted with small taps, for the removal of air, which would retard circulation if allowed to remain. Heating by hot water may be said to depend, in part, on the influence of gravity on water being to some extent overcome by heating in a boiler. It ascends the flow pipe by convection, where its onward journey would speedily end if it were not for the driving force of other molecules of water following, and the suction set up by the gravitation into the boiler of the cooled water by the return pipe. The power of water to conduct heat is very low. The conducting power of the iron in which it is conveyed is high. It is, however, probable that conduction is to some extent a factor in the process.

Pipes

It is a mistake to stint the quantity of piping, since it is far more economical and better for the plants to have a larger surface heated moderately than a smaller surface heated excessively. In view of the fact that air expands, becomes lighter and rises, under the influence of heat, the pipes should be set near the floor. If intended to raise the temperature of the structure, they should be set on iron or brick supports just clear of walls, earth or other heatabsorbing bodies. Those intended to provide bottom heat, however, are set in

(a) water tanks running under the beds, or

(b) in enclosed dry chambers under the beds, or are

(c) embedded in the soil or plunging material.

The first-named method is distinctly superior to the others. Pipes of 2 in., 3 in., 4 in. and 6 in. diameters are mostly used, the 4 in. size being the most convenient for general purposes. The joints are packed or caulked with tow, smeared with a mixture of white and red lead. Flanged joints are made to bolt together on washers of vulcanised rubber.

Boilers

There are numerous types of boilers in use, illustrative of efforts to secure as much exposure as possible to the action of the flames. The water-tube type, with multiple waterways, consists of a number of separate tubes j oined together in various ways. Some of these are built in the form of a blunt cone, and are known as conical tubular boilers. Others are built with the tubes arranged horizontally, and are known as horizontal tubular boilers.

The majority of the latter are more or less saddle-shaped. Boilers with a single waterway are of three principal types, the Cornish, the saddle and the conical. The Cornish is cylindrical with the furnace occupying about half the length of the cylinder. The saddle is so named from its supposed resemblance to a saddle. It is set to span the furnace, additional exposure to heat being secured in a variety of ways by flues. Exposure in the conical boiler is direct on its inner surface, and is supplemented by flues. Tubular boilers, especially the horizontal types, are very powerful and economical.

The Cornish type is a rather slow and steady boiler, and is much used for providing heat for large areas. The saddle boiler is very commonly employed to provide heat for moderately sized and small areas. Both are powerful and economical. Conical boilers are more expensive to set by reason of their shape, and are not so convenient to manipulate as the horizontal kinds.

Water supply—Wastage of water in the boilers should be made good automatically from a cistern controlled by means of a ball-cock. It should be placed as high above the boiler as practicable. The feed should connect with the return pipe near the point at which it enters the boiler.

Stokeholds

These have usually to be excavated to admit of the boilers being set below the level of the pipes they are intended to serve. In consequence of their depth, the draining of stokeholds often presents difficulties. Care should be taken to allow sufficient room to properly manipulate the fires and to store fuel. It is important that the ventilation should be as efficient as practicable, especially where coke fuel is to be used.

Stoking

The management of the furnaces is relatively easy, and consists in adapting the volume and intensity of the fires to particular needs. It involves the keeping clean of flues, ashpits and especially the fires themselves. Where coke or ordinary hard coal are used, the removal of clinkers should be done systematically, and the fires stirred. Anthracite coal fires should not be stirred more than is absolutely necessary, and should not be fed in driblets. They require more draught than coke fires, but care must be taken not to give too much, as excessive heat is likely to melt or soften the fire-bars. Draught is regulated in the ashpit by opening or closing the bottom door of the furnace and by the damper on the smoke shaft. The latter must be of a fairly good height, according to circumstances, to secure a good draught.

Solar Heat

The importance of sun heat to the general well-being of plant life, its influence on the production of flowers and the ripening of edible fruits, has long been appreciated in horticulture. The practice of "closing up" early in the afternoon, i.e. the closing of ventilators has for its object the conservation of as much solar heat as practicable.

Ventilation

This consists in the admission of air for the purpose of preventing stagnation of the atmosphere and for the regulation of temperature. Means of affording ventilation in all plant houses should be provided in at least two places—as near the floor as practicable, and at the top. Mechanical contrivances whereby whole sets of ventilators may be operated simultaneously are now in common use, and are much more convenient and economical than the older method of working each ventilator separately. Efficient ventilating can only be effected by the exercise of common sense and vigilance, and care must be taken to avoid cold draughts through the houses.

GARDEN MATERIALS AND APPLIANCES

The principal soils used in gardens, either alone, or mixed to form what are called composts, are—loam, sand, peat, leaf-mould and various mixtures and combinations of these made up to suit the different subjects under cultivation. Loam is the staple soil for the gardener; it is not only used extensively in the pure and simple state, but enters into most of the composts prepared specially for his plants. For garden purposes loam should be rather unctuous or soapy to the touch when moderately dry, not too clinging nor adhesive, and should readily crumble when a compressed handful is thrown on the ground. If it clings together closely it is too heavy and requires amelioration by the admixture of gritty material; if it has little or no cohesion when squeezed tightly in the hand, it is] too light, and needs to be improved by the addition of heavier or clayey material. Sound friable loam cut one sod deep from the surface of a pasture, and stacked up for twelve months in a heap or ridge, is invaluable to the gardener. When employed for making vine borders, loam of a somewhat heavier nature can be used with advantage, on account of the porous materials which should accompany it.

For stone fruits a calcareous loam is best; indeed, for these subjects a rich calcareous loam used in a pure and simple state cannot be surpassed. Somewhat heavy loam y are best for potting pine apples, for melons and strawberries, fruit trees in pots, and may be used with the addition of manures only; but for ornamental plants a loam of a somewhat freer texture is preferable and more pleasant to work. Loam which contains much red matter should be avoided. Sand is by itself of little value except for striking cuttings, for which purpose fine clean sharp silver sand is the best; and a somewhat coarser kind, if it is gritty, is to be preferred to the comminuted sands which contain a large proportion of earthy matter. River sand and the sharp grit washed up sometimes by the road side are excellent materials for laying around choice bulbs at planting time to prevent contact with earth which is perhaps manure-tainted. Sea sand may be advantageously used both for propagating purposes and for mixing in composts. For the growth of pot plants sand is an essential part of most composts, in order to give them the needful porosity to carry off all excess of moisture from the roots. If the finer earthy sands only are obtainable, they must be rendered sharper by washing away the earthy particles. Washed sand is best for all plants like heaths, which need a pure and lasting peaty compost.

Peat soil is largely employed for the culture of such plants as rhododendrons, azaleas, heaths. In districts where heather and gritty soil predominate, the peat soil is poor and unprofitable, but selections from both the heathy and the richer peat soils, collected with judgment, and stored in a dry part of the compost yard, are essential ingredients in the cultivation of many choice pot plants, such as the Cape heaths and many of the Australian plants. Many monocotyledons do well in peat, even if they do not absolutely require it. Leaf-mould is eminently suited for the growth of many freegrowing plants, especially when it has been mixed with stable manure and has been subjected to fermentation for the formation of hot beds. It any state most plants feed greedily upon it, and when pure or free from decaying wood or sticks it is a very safe ingredient in composts; but it is so liable to generate fungus, and the mycelium or spawn of certain fungi is so injurious to the roots of trees, attacking them if at all sickly or weakened by drought, that many cultivators prefer not' to mix leaf-mould with the soil used for permanent plants, as peaches or choice ornamental trees.

For quick growing plants, however, as for example most annuals cultivated in pots, such as balsams, cockscombs, globe-amaranths and the like, for cucumbers, and for young soft-wooded plants generally, it is exceedingly useful, both by preventing the consolidation of the soil and as a manure. The accumulations of light earth formed on the surface in woods where the leaves fall and decay annually are leaf-mould of the finest quality. Leaves collected in the autumn and stored in pits or heaps, and covered with a layer of soil, make beautiful leaf-mould at the end of about twelve months, if frequently drenched with water or rain during this period.

Composts are mixtures of the foregoing ingredients in varying proportions, and in combination with manures if necessary, so as to suit particular plants or classes of plants. The chief point to be borne in mind in making these mixtures is not to combine in the same compost any bodies that are antagonistic in their nature, as for example lime and ammonia. In making up composts for pot plants, the fibrous portion should not be removed by sifting, except for small-sized pots, but the turfy portions should be broken up by hand and distributed in smaller or larger lumps throughout the mass. When sifting is had recourse to, the fibrous matter should be rubbed through the meshes of the sieve along with the earthy particles. Before being used the turfy ingredients of composts should lie together in a heap only long enough for the roots of the herbage to die, not to decompose.

Manures—These are of two classes, organic and inorganic—the former being of animal and vegetable, the latter of mineral origin. The following are organic manures: Farm-yard manure consists of the mixed dung of horses and cattle thrown together, and more or less soaked with liquid drainings of the stable or byre. It is no doubt the finest stimulant for the growth of plants, and that most adapted to restore the fertile elements which the plants have abstracted from exhausted soils. This manure is best fitted for garden use when in a moderately fermented state.

Horse dung is generally the principal ingredient in all hot bed manure; and, in its partially decomposed state, as afforded by exhausted hot beds, it is well adapted for garden use. It is most beneficial on cold stiff soils. It should not be allowed to lie too long unmoved when fresh, as it will then heat violently, and the ammonia is thus driven off. To avoid this, it should be turned over two or three times if practicable, and well moistened—preferably with farm-yard drainings.

Cow dung is less fertilising than horse dung, but being slower in its action it is more durable; it is also cooler, and therefore better for hot dry sandy soils. Thoroughly decayed, it is one of the best of all manures for mixing in composts for florists' flowers and other choice plants. Pig dung is very powerful, containing more nitrogen than horse dung; it is therefore desirable that it should undergo moderate fermentation, which will be secured by mixing it with litter and a portion of earth. When weeds are thrown to the pigs, this fermentation becomes specially desirable to kill their seeds.

Night-soil is an excellent manure for all bulky crops, but requires to be mixed with earth or peat, or coal-ashes, so as both to deodorise it and to ensure its being equally distributed. Quicklime should not be used, as it dispels the greater part of the ammonia. When prepared by drying and mixing with various substances, night-soil is sold as desiccated night-soil or native guano, the value of which depends upon the materials used for admixture.

Malt-dust is an active manure frequently used as a top-dressing, especially for fruit trees in pots. It is rapid in its action, but its effects are not very permanent. Rape dust is somewhat similar in its character and action.

Bones are employed as a manure with decided advantage both to vegetable crops and to fruit trees, as well as to flowers. For turnips bone manure is invaluable. The effects of bones are no doubt mainly due to the phosphates they contain, and they are most effectual on dry soils. They are most quickly available when dissolved in sulphuric acid.

Guano is a valuable manure now much employed, and may be applied to almost every kind of crop with decided advantage. It should be mixed with six or eight times its weight of loam or ashes, charred peat, charcoal-dust or some earthy matter, before it is applied to the soil, as from its causticity it is otherwise not unlikely to kill or injure the plants to which it is administered. Peruvian guano is obtained from the excreta of South American sea-birds, and fish guano from the waste of fish. Both are remarkable for the quantity of nitrates and phosphates they contain.

Pigeon dung approaches guano in its power as manure. It should be laid up in ridges of good loamy soil in alternate layers to form a compost, which becomes a valuable stimulant for any very choice subjects if cautiously used. The dung of the domestic fowl is very similar in character.

Horn, hoof-parings, woollen rags, fish, blubber and blood, after treatment with sulphuric acid, are all good manures, and should be utilised if readily obtainable. Liquid manure, consisting of the drainings of dung-heaps, stables, cowsheds, or of urine collected from dwelling houses or other sources, is a most valuable and powerful stimulant, and can be readily applied to the roots of growing plants. The urine should be allowed to putrefy, as in its decomposition a large amount of ammonia is formed, which should then be fixed by sulphuric acid or gypsum; or it may be applied to the growing crops after being freely diluted with water or absorbed in a compost heap.

Liquid manures can be readily made from most of the solid manures when required, simply by admixture with water. When thus artificially compounded, unless for immediate use, they should be made strong for convenience of storage, and applied as required much diluted. The inorganic manures are following: Ammonia is the most powerful and one of the most important of the constituents of manures generally, since it is the chief source whence plants derive their nitrogen. It is largely supplied in all the most fertilising of organic manures, but when required in the inorganic state must be obtained from some of the salts of ammonia, as the sulphate, the muriate or the phosphate, all of which, being extremely energetic, require to be used with great caution. These salts of ammonia may be used at the rate of from 2 to 3 cwt. per acre as a top-dressing in moist weather. When dissolved in water they form active liquid manures. The most commonly used nitrogenous manures are nitrate of soda, nitrate of potash and sulphate of ammonia, the prices of which are constantly fluctuating.

Potash and soda are also valuable inorganic manures in the form of carbonates, sulphates, silicates and phosphates, but the most valuable is the nitrate of potash. The price, however, is generally so high that its use is practically nil, except in small doses as a liquid manure for choice pot plants. Cheaper substitutes, however, are now found in sulphate of potash, and muriate of potash and kainit. The two last-named must not be applied direct to growing crops, but to the soil some weeks in advance of sowing or cropping. The manures of this class are of course of value only in cases where the soil is naturally deficient in them. On this account the salts of soda are of less importance than those of potash. The value of wood ashes as a manure very much depends upon the carbonate and other salts of potash which they contain.

Phosphoric acid, in the form of phosphates, is a most valuable plant food, and is absorbed by most plants in fairly large quantities from the soil. It induces the earlier production of flowers and fruits. In a natural state it is obtained from bones, guano and wood ashes; and in an artificial condition from basic slag or Thomas's phosphate, coprolites and superphosphate of lime. Lime in the caustic state is beneficially applied to soils which contain an excess of inert vegetable matter, and hence may be used for the improvement of old garden soils saturated with humus, or of peaty soils not thoroughly reclaimed. It does not supply the place of organic manures, but only renders that which is present available for the nourishment of the plants. It also improves the texture of clay soils.

Gypsum, or sulphate of lime, applied as a top-dressing at the rate of 2 to 3 cwt. per acre, has been found to yield good results, especially on light soils. It is also employed in the case of liquid manures to fix the ammonia. Gas lime, after it has been exposed to the air for a few months is an excellent manure on heavy soils. In a fresh state it is poisonous and fatal to vegetation, and is often used for this reason to dress land infested with wireworms, grubs, club-root fungus.

Burnt clay has a very beneficial effect on clay land by improving its texture and rendering soluble the alkaline substances it contains. The clay should be only slightly burnt, so as to make it crumble down readily; in fact, the fire should not be allowed to break through, but should be constantly repressed by the addition of material. The burning should be effected when the soil is dry.

Vegetable refuse of all kinds, when smother-burned in a similar way, becomes a valuable mechanical improver of the soil; but the preferable course is to decompose it in a heap with quicklime and layers of earth, converting it into leaf-mould. Potato haulms, and club-rooted cabbage crops should, however, never be mixed with ordinary clean vegetable refuse, as they would be most likely to perpetuate the terrible diseases to which they are subject. The refuse of such plants should be burned as early as possible. The ash may be used as manure. Soot forms a good top-dressing; it consists principally of charcoal, but contains ammonia and a smaller proportion of phosphates and potash, whence its value as a manure is derived. It should be kept dry until required for use. It may also be used beneficially in preventing the attacks of insects, such as the onion gnat and turnip fly, by dusting the plants or dressing the ground with it.

Common salt acts as a manure when used in moderate quantities, but in strong doses is injurious to vegetation. It suits many of the esculent crops, as onions, beans, cabbages, carrots, beet-root, asparagus, the quantity applied varies from 5 to to bushels per acre. It is used as a top-dressing sown by the hand.

REFERENCES

Adams, C K, 2008. *Principles of Horticulture.* Butterworth-Heinemann.

ITC. 1988. *Manual on the packaging of fresh fruits and vegetables.* International Trade Centre, UNCTAD/GATT, Geneva.

Kader, A. et al., 1985. *Post-harvest technology of horticultural crops.* Cooperative extension, University of California

McGee, J.R. and Kruse, M. 1986. "Swidden horticulture among the Lacandon Maya". University of California, Berkeley: Extension Media Center,.

Sargent, M.J. 1973. *Economics in horticulture,* Macmillan Press.

Thompson, S.I. 1977. "Women, Horticulture, and Society in Tropical America". *American Anthropologist,* N.S., 79: 908-910

10

Nursery Management

A nursery can be part of a diversification strategy to make a farm more profitable or a nursery can be a sole enterprise. In either case, it is important to start small and expand later. Sustainable nursery practices aim to reduce levels of synthetic fertilisers and pesticides, use integrated pest management systems to deal with insects, diseases and weeds and focus on building the soil to promote plant health.

The most important things to consider before starting production are what crops to grow and how to market them. In today's economy, it is no longer possible to grow crops without first considering the crop's marketability. Here are some facts to keep in mind before starting out:

- Container-grown crops generate about 10 times more sales per acre than field crops.
- Lawn and garden centers draw approximately 80 percent of their customers from a 5- to 15-mile radius. More than 60 percent of an average wholesale nursery's sales are to customers from within the state. Small nurseries sell about 20 percent of their plants to out-of state customers.
- Retail garden centers usually want small plants in 1- to 3-gallon containers. Landscaping firms and landscapers want larger container plants in 3- to 5-gallon containers and balled and burlapped woody plants.
- The nursery industry is dependent on the construction industry and on the rate of unemployment in the vicinity of the nursery.

GENERAL PRODUCTION

There are two types of nursery production: *field* and *container*. Field stock is either direct-seeded or transplanted from seedlings and then lifted as bare-root stock for use as nursery liners, fruit trees, seedlings for Christmas trees, windbreaks and conservation plantings. Field stock is also grown for balled and burlapped landscape or shade trees.

Container stock, which is propagated from seed, rooted cuttings and field-grown seedlings, is common in both forestry and landscape nursery production. Fifty years ago, nursery managers grew ornamental plants in the field and dug the plants up later for transplanting. Today, 80 percent of ornamental plants are container grown. The switch occurred for several reasons. Container-grown trees have a greater chance for survival and establishment after transplanting. Containerised production does not require good soil and takes up less acreage. Containerised stock also enables the grower to extend the planting season.

Container and field production will be discussed separately, but there is commonality between the two forms of production. Most woody landscape plants, regardless of how they are grown, are propagated by cuttings. Both types of production require spending a high percentage of a nursery's budgets on farm-type mechanised implements and fertilisers.

Soil productivity is not as important when growing only containerised products, but relatively level land with good drainage is still necessary. Beginning nursery managers must learn the length of time required to produce marketable crops and how to schedule planting so the proper number of each species is available for the first year of sale and following years.

IRRIGATION MANAGEMENT

The two most widely used irrigation systems are overhead and drip or trickle systems. Overhead irrigation systems are designed to cover a large area and these systems are the least expensive to install. However, this method produces uneven water distribution, which can slow plant growth, encourage disease and contribute to runoff. A container nursery using overhead irrigation can use from 15,000 to 40,000 gallons of water per acre daily in the summer, a reminder that sufficient water is a prerequisite to nursery production.

Large containers are usually watered with a drip or trickle system, which uses 60 to 70 percent less water than an overhead system. Drip irrigation systems cost more to install but have superior application uniformity and efficiency. Drip irrigation systems are also affected less by wind and crop canopies and produce less runoff. Another advantage is that workers can continue working while the plants are being irrigated. The biggest disadvantage to a drip or trickle irrigation system, besides the initial cost, is keeping the pipes and emitters clean.

A third, less-used type of irrigation system is subirrigation using capillary sandbeds. In this system, water rises into containerised plants through capillary action. The sandbed is covered with at least 1 inch of fine sand and slopes very slightly. Water is released at the high end and slowly percolates to the low end. These systems cost the most to install, but have no runoff or leaching.

Sandbeds are normally built using wood sidewalls, a plastic bed liner, sand, a small tank, a drainpipe and a float valve. Sandbeds do not require the use of any electrical parts and provide a uniform and consistent supply of water without forming a saturated water table at the base of the soil column in the container. Sandbeds offer efficient and uniform crop growth while providing less water, less fertiliser and less pesticide. Sandbeds also require less labour since sprinkler heads, timers, pumps, valves and water-treatment systems don't need to be monitored.

The biggest disadvantage of sandbeds is that weeds and containerised plants grow into the structure. The Agroliner is a retail product designed to alleviate this problem. The Argoliner is a mat treated with Spin Out, a copper paint that prevents root growth and is registered by the Environmental Protection Agency. The mat is placed over the sand and under the containers.

Plants need to be watered often, especially during hot, sunny days. A typical nursery plant in a 1-gallon container can consume a pint of water a day, while the growing medium capacity may be only 1.5 pints. One important aspect of irrigation management is to group plants according to water requirements.

The most important issue with irrigation in sustainable nursery production is water, fertiliser and pesticide runoff. Many states now have regulations limiting runoff and groundwater nitrate levels.

Subirrigation systems are designed to eliminate runoff, but overhead and drip or trickle systems may require special attention. Ditches planted with grass to slow down water flow or tile systems that direct water to a pond or other holding area can collect runoff water.

The water and some of the fertilisers present can be recycled by pumping water back out of the holding tank or pond after impurities like sand and silt settle out. Recycled water can improve plant growth. In experiments with more than 100 species of ornamentals grown in 2.8-liter containers, the mean relative growth of plants irrigated with continuously recycled water was 103 percent of the control.

Pulse irrigation is another way to reduce runoff. In this system, a small amount of water is applied five or six times a day, instead of one heavy watering daily. Very little water escapes from the container or runs off from the field. Less fertiliser is applied because there is less leaching. Most nurseries that use this system use a computer to control water flow, since watering plants repeatedly by hand causes a huge increase in labour expenses. There are several cultural practices that can reduce runoff:

- Avoid irrigating bare soil
- Have rough soil surfaces to provide surface storage of water
- Use less-porous media that retain moisture and nutrients
- Use slow-release fertilisers instead of liquid fertilisers

Researchers at The Ohio State University conducted experiments to reduce the amount of pesticides and growth regulators leached from nursery pots and trays. Researchers had excellent success mixing chemicals in ordinary latex paint and then painting the interior of the pots. The pots leached less and the growth regulator and pesticide researchers used, commercially available brands Bonzi and Marathon, provided more consistent control. This method also reduced worker re-entry intervals in the nursery area, since workers apply chemicals once at the beginning of the growth process.

Reducing moisture stress enhances growth more than increasing fertiliser concentration, a study conducted in the late 1990s found. This study, also conducted at The Ohio State University, used fertiliser concentrations between 50 and 200 milligrams per liter of nitrogen. The researchers showed that water stress might limit growth more frequently than limited nutrition under current container production practices. The researchers recommend using lower amounts of N fertiliser—about 50 milligrams per liter—and providing sufficient moisture.

PEST MANAGEMENT

Integrated pest management is an ecologically based pest control strategy that is part of the overall crop production system. It is called integrated because all appropriate methods from multiple scientific disciplines are combined into a systematic approach for optimising pest control. Management implies acceptance of pests as inevitable components at some population level in the agricultural system.

An integrated pest management program involves using resistant cultivars, building up populations of beneficial organisms, monitoring numbers of pests, developing treatment thresholds and using spot treatments of pesticides that are the least harmful to beneficial organisms and the environment. It is important to identify pests early so nursery managers can take appropriate measures quickly.

CONTAINER PRODUCTION

A wide selection of ornamentals is produced in containers. Homeowners usually prefer to buy containerised plants because the plants are easier to transport and transplant than balled and burlapped plants. The following section summarises some important container product ion practices and addresses sustainable nursery management issues like recycling plastics, weed control and fertilisation. The advantages of containerised production include:

- Achieving high plant densities
- Using land unsuited for field production
- Planting at times independent of the weather
- Eliminating some operations, like root pruning
- Lowering transportation costs because of lightweight media
- Experiencing less root loss and a greater chance of survival than with field-grown trees

The disadvantages are also numerous:

- Small containers need frequent watering
- Nutrients deplete rapidly
- Plants require winter protection
- Plants easily become root-bound
- Wind can knock over trees
- Containers are costly
- Labour costs to pot up plants are high
- Temperature extremes stress roots

Growing certified organic nursery stock intended for sale to vineyards, berry farms and orchards raising organic produce is a niche market that requires special attention. Restricted products include common ingredients in conventional nursery production like chemical fertilisers, wetting agents, herbicides and synthetic insecticides and fungicides.

There are several factors to keep in mind when deciding what containers to use. Factors include cost, design features that control root growth, durability, shipping capacity, availability, how the container affects growing medium moisture content and temperature and how the container suits the particular needs of the nursery.

Round, black plastic pots are the industry standard, but can cause root constriction that leads to plants with poorly developed root systems. There are other kinds of containers that promote better root systems. Copper-lined, white and light-colored containers all produce more root growth and square and stair-step pots help keep plants from becoming root-bound. Pots and containers designed for enhanced root growth are an important feature in containerised nursery production. Each pot and container offers its own advantages and disadvantages.

Foresters discovered that copper can control root growth. Copper kills root tips that come in contact with it, forcing roots to branch within the root ball instead of circling around it.

Plants grown in copper-treated containers are taller, less root-bound and have higher transplant survival rates. Plants grown in copper-treated containers also have increased nitrogen recovery and require fewer applications of nitrogenous fertiliser. More than 120 species perform better in copper-treated containers versus untreated containers. Copper-treated pots also do not leach or leach very little into groundwater or soil.

Copper-treated fiber pots, made from recycled paper, are biodegradable and can even be composted. The main problem with fiber pots is that the pots can degrade too quickly. Research at The Ohio State University showed that incorporating copper into fiber pots can increase their longevity.

Bottomless Pots

Air root pruning is another way to prevent root circling. Air root pruning employs a similar mechanism to copper-treated pots. Root tips that come in contact with air are killed and the root system branches out within the root ball.

Growing tree seedlings in bottomless paperbased milk cartons is one way to use air root pruning. The milk carton, when folded open, creates a long, bottomless container. These containers are placed in a wooden flat with a wire-screen bottom and then filled with a soilless nursery mix. Another option is to place tree seedlings in plastic milk crates. The taproot grows downward and out through the bottom of the container. The root tip is exposed to the air, desiccates and dies back.

Repeated air-root pruning stimulates lateral branching and results in a fibrous root system as opposed to a strong taproot system. The benefit to the tree is rapid establishment in the field or landscape with increased scaffold branching and top growth. Nursery stock production by the milk carton method is especially useful for on-farm tree production and can be used in the propagation of a wide range of woody plants, including strong tap-rooted species such as black walnut, pecan and pines for Christmas trees. There are other types of containers that promote excellent root branching and discourage root circling. Root-Maker pots have staggered walls and a staggered bottom, which prevent root circling and direct roots toward holes in the walls and bottom of the pots. Whitcomb, formerly head of the nursery research program at Oklahoma State University, is well known for his numerous innovative approaches to unusual container systems.

Tubes

Long bottomless tubes are another production system that uses air root pruning. Tubes are generally made of plastic or Styrofoam. Nurserymen can use single tubes or several tubes imbedded in a flat. Tube plants range in size from large plugs sold as nursery liners to seedling trees grown in long, narrow pots and sold directly to consumers. Tubes are popular because they allow massive plant quantities to grow in a small area. Tubes are particularly adaptable to small-scale nursery production and to specialised stock like perennials and tree seedlings.

Pot-in-pot system

The pot-in-pot production method alleviates some of the problems associated with container production, such as blow-over and moisture loss. This system involves burying a holder pot, or moat pot, in the ground and placing a containerised plant inside this pot. The main drawback to this system is the high initial cost of the moat pot. The moat pot is a long-term investment since it will last 15 years or more.

Plastic Recycling

Most nurseries use lots of plastic in the form of pots, flats, hanging baskets and greenhouse film. The nursery can reuse some of these products, but it's important to have a recycling system in place. Buying multiyear, ultraviolet-stabilised greenhouse film decreases the amount of sheet plastic used each year, but this kind of film is very expensive and not always readily available. Fortunately, there are a number of recyclers around the country that accept nursery plastic.

Many recyclers require nurseries to gather a certain amount of plastic before sending a truck to pick it up. Smaller nurseries may have trouble storing a large amount of plastic. Try combining plastic waste with other growers in the community. Some recyclers will not pay the grower for the plastic but do not charge for transportation costs, which are often high.

Weed Control

Weed control is extremely important in container production. Weeds compete for water and nutrients and hinder sales of nursery stock. Weed control efforts should focus on two areas: in the pot and under the pot. Sanitation is the least costly and most effective method for controlling weeds. To prevent weed seeds from blowing into pots, a vegetation-free zone on and surrounding the production bed is critical. To keep weeds from growing under the pot, place pots on geotextile weed barriers, often called fabric weed barrier or landscape cloth. Modern landscape cloths are durable and can last for

10 to 12 years in full sun. Landscape cloths do an excellent job of controlling weeds and are permeable to water from irrigation and rainfall, so drainage is not a problem. Although the initial cost is high, the expense can be prorated as an annual weed control investment.

Hand weeding is costly, but it may be appropriate in a small nursery setting. Weeds must be removed when they are small since large quantities of media are lost when pulling big weeds out of containers. Herbicides are widely used in container nursery production. Growers use a weedfree medium to establish nursery plants, but wind, birds and surface irrigation water all deposit weed seeds onto the pot surface. Broadleaf and grassy weeds love to get a free ride into a container nursery because growing conditions in a media-rich pot are perfect. Pre- and post-emergent herbicides are commonly used in commercial nursery production to control these opportunists.

Fabric weed barrier disks can control weeds in containers. The fabric disks are pre-slit and fit on top of the pot around the plant stem. Barrier disks prevent weeds from growing in containers by excluding sunlight and inhibiting weed germination. The disks are permeable to air and water but prevent germination of troublesome container nursery weeds like oxalis. The disks also reduce evaporation.

Tex-R Geodiscs are fabric disks treated with Spin Out. Geodiscs prevent weed growth by excluding light and pruning the roots of weed seeds that land on the fabric. Geodiscs provide effective weed control for up to three years and can be moved from pot to pot. Bonnie Appleton, director of the horticulture master's degree program at Virginia Tech, recently conducted research using Geodiscs on container-grown willow oaks. The Geodiscs suppressed all weeds completely. Trees grown in the pots with Geodiscs had higher top dry weights and root dry weights than trees grown without any form of weed control and trees sprayed with a conventional herbicide.

Corn gluten meal, a recently introduced weed control, is a by-product of corn syrup processing. Corn gluten meal is a pre-emergent herbicide applied in early spring. The meal works best when applied to the top one-quarter inch of soil and must be reapplied every year. Corn gluten meal is 10 percent nitrogen and acts as a slow-release fertiliser for the crop. Corn gluten meal is patented and sold as an herbicide.

There are some new environmentally friendly contact herbicides that break down quickly and provide options for weed control in container nurseries around irrigation risers and perimeter areas and are also applicable for general use in field nursery production. One class of products is made from pelargonic acid, a fatty acid found in plants and animals. The products are sprayed on weeds and rapidly lower the weeds' pH level, weakening cell walls and killing the weeds within two hours. A second class of products contain acetic acid (vinegar), lemon juice, eugengol, thyme oil, orange oil and other natural ingredients. The products work as contact herbicides and control, with varying degrees of success, broadleaf and grassy weeds. Application to nursery plants should be avoided and several applications may be necessary to kill perennial weeds.

Fertilisation

Large-scale container nursery production is a huge success largely due to advances in media and fertiliser combinations. Favorable media and fertiliser combinations are a result of several decades of research collaborations between land-grant universities, commercial nurseries and the fertiliser industry. Commercial synthetic fertilisers including slow-release and liquid fertilisers play a key role in this picture. As organic production becomes standardised under the National Organic Program Final Rule, more nursery growers explore fertilisers acceptable in organic production. Unlike synthetic greenhouse fertilisers, there is minimal research to support the use of organic fertilisers in a nursery mix recipe. Most of the following material will focus on organic fertilisers for container nursery production. In sustainable nursery production the emphasis is eliminating runoff, regardless of if the fertiliser source is synthetic or organic. Excessive nitrates and phosphorus are the most common problems in runoff water.

There are four basic ways to fertilise containerised plants: incorporate, topdress, liquid feed and foliar feed. In a nursery container, fertiliser incorporation in the nursery mix combined with liquid feeding should provide sufficient nutrition. Several organic fertilisers provide nitrogen. Fertilisers include alfalfa meal, blood meal and cottonseed meal, among others. Materials that provide phosphorus include oak leaves, bone meal and shrimp wastes, among others. Greens and, granite meal and soybean meal all provide potassium.

Maintaining adequate levels of nutrients in the container medium is necessary for optimum growth of woody ornamentals. The levels of soluble nutrients in containers can be significantly reduced after three or four irrigations because of limited container volume and frequent application of water. Use slow-release and liquid fertilising systems to overcome this problem. Organic or synthetic slow-release fertilisers help cut down levels of nitrates in runoff water. Slow-release and controlled-release synthetic fertilisers, like commercially available Nitroform and Osmocote, are common in container production systems. Incorporate slowrelease and controlled-release fertilisers into the growing media for best results. Do not topdress.

Slow-release fertilisers are often used in combination with liquid fertilisation. Nitrogen is the main nutrient supplied through liquid feeding, or fertigation. Organic liquid fertilisers include fish emulsion, fish powder, blood meal, bat guano, seabird guano, worm castings and composted manure teas. Some forms of organic fertilisers are more adaptable to low-volume irrigation systems like drip or trickle systems. Somes found that spray-dried fish protein and poultry protein fertilisers do not clog drip emitters and microsprinklers. Fish protein, blood protein, poultry protein and brewers yeast are all available as spray-dried materials.

Foliar feeding can supplement soil and liquid fertilisation, especially where certain nutrients are deficient and must be incorporated into the plant quickly. Use filtered solutions of manure, seaweed, fish powder and fish emulsion. Seaweed is an excellent foliar material because it contains growth hormones, including auxins, gibberellins and cytokinins, as well as trace elements. Research suggests that foliar feeding programs enhance plant resistance to pest and disease attack. Compost teas are popular as a foliar feed primarily because of their disease-suppressive characteristics.

Potting Media

Field soil is sometimes used in container mixes. Field soil makes up 10 to 30 percent of the mix by volume, but soil is heavy and requires the additional step of pasteurisation to eliminate diseases and weed seeds. The standard replacement for soil is peat moss, but there is concern that peat is a non-renewable resource. Research is being conducted to determine what materials can be used to replace peat. Most of the products being tested are some form of waste.

Composted pine bark, a by-product of the lumber industry, is an excellent medium for containerised plants. Mixes containing more than 20 percent composted pine bark support a significant level of suppression of *Pythium* damping-off. Other alternatives are coir, spent mushroom compost, paper mill sludge, apple pomace, shredded newspaper, compost, processed alfalfa, processed kenaf, recycled cardboard and composted municipal yard waste. These alternative products should not compose more than 50 percent of the mix.

Mycorrhizae are soil fungi that form beneficial associations with plant roots. Mycorrhizae enable plant roots to do a better job of gaining nutrients and water. The fungi can be used in field or container production. Growers achieve better stand establishment, use less fertiliser and inoculate bare-root seedlings when using mycorrhizae. Commercially available mycorrhizae stimulate the roots of almost all tree and shrub species.

FIELD PRODUCTION

Until the 1950s, virtually all nursery production occurred in the field. Field production is still widely used to produce bare-root seedlings for conservation plantings, fruit trees and nursery liners. The most profitable product of field nurseries is bagged and burlapped shade trees for the landscape industry. In-ground production is advantageous to tap-rooted tree species, mass plantings, inexpensive establishment and large caliper size. Disadvantages of in-ground production include a higher percentage of plant loss and longer establishment periods after transplanting. Field nursery production involves the use of unique soil management practices. Soilbuilding cover crops and crop rotations are important to maintain good soil structure, fertility and organic matter. Living mulches are cover crops planted in the aisles to hold the soil, provide traction, increase water infiltration and suppress weeds. Legume cover crops fix nitrogen and can be used to reduce the amount of nitrogen fertiliser applied each year.

Integrating living mulches, cover crops and the application of high-quality composts in field nursery operations are the fastest ways to improve nursery soil. To reduce pest problems, plant a diversity of species rather than a large block of single species followed by a large block of another species. Habitat management for beneficial insects is also an option in a field situation.

Nursery equipment and irrigation systems for field nursery production are unique. A recent innovation in field nursery production is the use of in-ground fabric containers, sometimes called root control bags or field grow bags. The bags have a fabric or clear polyethylene bottom stitched or glued to walls made of non-woven fabric and come in several sizes. In theory, the bags combine the best qualities of container and field production.

The advantages of field grow bags are numerous. Bags enhance rooting, leave 80 percent of roots intact during transplanting, make harvesting easier, save labour and time, do not require special machinery at harvest and can be harvested year-round. Bagged and burlapped plants are normally harvested only during dormancy.

There are also disadvantages. The initial investment in grow bags is expensive, plants need more staking and water after transplanting, damaged bags cannot be used, mechanical cultivation and precise fertiliser application are difficult and bag removal can be difficult and time consuming. The greatest disadvantage of growing in bags is marketing the plants. Few people know the advantages of growing in bags and find bagged trees less convenient to handle than containerised trees. One way around this is to grow the plant in a pot for the last year of production.

An important consideration in bagged and burlapped production is the loss of 200 to 250 tons of topsoil per acre at each harvest. The digging and removal of topsoil from bagged and burlapped nursery operations is a practice that can, over a long time, seriously deplete the farm's most important resource. There are two ways to fight topsoil depletion. Replace topsoil with something else or have a bare-root operation that does not require topsoil to leave the farm. Many growers use compost to replace some of the topsoil that is lost, but applying more than 40 tons of compost per acre is not recommended.

Weed Management

Weed control in modern field nursery production is based on the use of herbicides. There are many excellent non-chemical alternatives, however. These include mechanical cultivation, flame weeding, mulches, living mulches, steam and solarisation.

Flame Weeding

Flame torches, or flamers, may be an option in some nursery situations. Flaming works by searing and disrupting plant cells, not by burning plant tissue. Passing a flamer quickly over a weed is enough to kill the top of the weeds, but roots can re-sprout new growth. Broadleaf weeds are more susceptible to flaming than grassy weeds. Flaming needs to be repeated every two to three weeks to control grasses.

Flame weeders can be used to prepare a stale seedbed by flaming off the first and second flushes of weeds to emerge after seedbed preparation. Weed flamers can also control post-emergent weeds. To protect young seedlings from injury, use flaming shields. Taller seedlings and trees with well-developed bark can withstand directed flaming aimed at weeds growing in and between the rows. Although there is some criticism that flaming is not a sustainable practice because it uses fossil fuels, flamekilling a nursery bed or field of seedlings uses less fossil fuel than manufacturing, transporting and spraying an herbicide for the same job.

Mulches

Mulches are another way to exclude weeds. Mulches keep out weeds by limiting light and retaining moisture in the soil. Organic mulches should be 3 to 4 inches thick and need replenishing once or twice a year. Millcreek Manufacturing, based in Leola, Pa., offers a row mulching machine that can apply mulch and compost to fieldgrown stock. The machine costs about $5,000 and can mulch beds from 18 to 48 inches wide, from ½ inch to 10 inches deep.

Landscape fabric can also be used in field production. A fast way for growers to get into production is to lay cloth in the field, cut or burn holes in the cloth and then plant the liners or seedlings. Researchers at Oregon State University found that mulches made of oyster shell, hazelnut shell and copper-treated geotextiles provide good suppression of liverwort, a prevalent weed in many nurseries. These three mulches outperformed mulches of rockwool, peat moss, coarse sand, perlite, pumice and the commercial herbicide Ronstarand Surfl an.

Living mulches

In a 1990s study conducted in Minnesota, researchers compared soil cultivation, herbicides and three living mulches for weed suppression in a field with six species of ornamental trees. The researchers used Norcen bird's-foot trefoil, Wheeler winter rye and a grass sod consisting of 80 percent Eton perennial ryegrass and 20 percent Ruby red fescue. The grass sod provided excellent weed control, but was overly competitive with the trees. The trefoil was infested with weeds. The winter rye, which was killed with herbicides and then acted as a mulch, provided good weed control and increased water infiltration and soil moisture. That evened out soil temperature fluctuations, reduced soil bulk density, improved nutrient cycling and reduced field maintenance costs. In general, the cover crops tended to reduce annual weeds and favour perennial species.

Steam

For years, conventional production systems used methyl bromide to sterilise soil before planting. One sustainable system that yields the same results uses steam to disinfest beds and greenhouses prior to planting. In a field planting, this system can treat planting beds. Steam is nontoxic, easy to apply, controls the same soil pests as conventional methods and works in a wide variety of climates and conditions.

A small portable boiler is the best unit to use for a greenhouse. The important differences in steam machines are how much heat the machines put out, how portable the machines are and how far into the soil the steam penetrates. Although most machines heat only the top 3 to 6 inches of soil, temperatures are high enough to kill most weed seeds. Machines that heat the soil to 140 degrees for at least 30 minutes kill pest fungi, bacteria, nematodes and weed seeds.

Solarisation

Soil solarisation is another option for killing pests before planting trees, shrubs or perennials. Treat only beds that will be planted. The basic principle of solarisation involves stretching sheets of clear plastic across moist ground. Solar radiation heats the soil and kills pests, including weed seeds and harmful insects. Solarisation can kill annual and perennial weeds if summer temperatures climb high enough. Solarisation can also be used to disinfest reused or soil-based potting media. To do this, enclose the media in plastic bags and leave them in the sun for two to three weeks. Two layers of plastic kill more pests and work about four times faster than one layer.

Harvest and Storage

At some point, nurseries must dig up fieldgrown trees and plants. Plants are often stored after digging. Conventional production systems dig trees in late fall or early winter and store trees in warehouses until early spring. During this time, nurseries spray bare-root trees with fungicides and bacteriostats to keep problems from arising. Researchers in Rhode Island experimented with Taxus bagged and burlapped stock to prevent the plants from rooting out, a condition where the roots grow into the burlap bags. They used Spin Out, an EPA-approved copper paint, in several different ways. Treatments included painting the bottom of the root ball with copper paint, setting the root ball on copper-treated burlap and rewrapping the root ball with copper-treated burlap before mulching. Although all these treatments provided good control of rooting out after 12 to 16 weeks, the most effective treatments were setting the root ball on copper-treated burlap and leaving it unmulched. The researchers also found that placing the root balls on TexR Agroliner, a Spin Out-treated non-woven fabric, stopped rooting out completely.

COSTS OF NURSERY PRODUCTION

The costs of nursery production include overhead, direct and marketing costs. Overhead costs include all the general costs of operating the nursery, like taxes, depreciation, interest, rent, utilities, insurance, maintenance and repair, new construction, new equipment, supplies, managerial and administrative salaries and labour wages that cannot be assigned to a particular crop. Direct costs are tied to a crop, like purchasing seed, potting media and fertilisers. Keeping excellent records is the best way to accurately determine true costs. Prices should reflect:

- Exact production costs that include a reasonable profit for each crop Prices and quantities offered by competitors
- Supply and demand for the crop, except prices for high-quality products and loyal customers

Competitive prices can fall below the cost of production. In this circumstance, a nursery with a unique advantage, like proximity to its market or a superior product, may be able to maintain a higher price that covers costs without experiencing a serious drop in the number of plants sold.

Although it is desirable to make a profit on each kind of plant, sometimes it is good marketing strategy to grow some plants that may not be profitable in order to offer a well-rounded inventory. A small nursery might specialise in a few high-quality plants or produce some plants not carried by larger nurseries, which produce only plants with high sales volumes.

One way to cut production costs is to grow plants in smaller containers. Although the crop sells for less, the costs of media and containers are reduced, as well as the time needed to produce the crop. Selling plants at wholesale prices means less money received for each plant, but also less money and time spent on marketing and advertising.

MARKET ANALYSIS

A market analysis is necessary to determine what opportunities exist to sell plant materials in the local area. Most new firms begin with only a few acres of production and market in a 50-mile radius, unless growing for mail order or on contract. A market analysis includes finding out what crops other nurseries grow successfully in the region. The analysis also evaluates competition potential from area nurserymen.

Marketing starts with a decision about what to produce and at what volume. Nursery managers need to:

1) Determine what kind of customer the nursery will attract and what size of plants those customers want.

   - Mass merchandisers want large volumes of a few popular plant species. Mass merchandisers purchase smaller plants. These customers may not care about buying specific plants, but focus on obtaining a good mix of fast-moving materials. Demand from these customers is seasonal. Plants in fashion vary from year to year.

     There are several disadvantages to dealing with mass merchandisers. These customers want instant shipment, pay the lowest price for plants and often do not take care of plants after receiving them, which can reflect poorly on a nursery.

   - Landscapers look for large, high-quality specimens carefully identified by cultivar. Landscapers generally want to buy plants from a limited number of producers but also want to choose among many plants and plant sizes. Landscaper purchases are spread out through the year, with an emphasis on spring planting. Horticulture degrees with specialisation in landscape architecture were popular career choices in the 1980s and 1990s, creating a solid base for nursery customers.

   - Lawn and garden centers fall somewhere between the mass merchandiser and the landscaper. Some centers want variety in plants and plant sizes, some do not.

     Other retail outlets include mail order, Web sites, farmers' markets and starting a landscaping business. Sales at farmers' markets will be local, but local can mean weekly travel of 200 to 300 miles to a large metropolitan area. Potential customers all share a common need to get uniform, well-grown plants from a producer without having to inspect the crop before each purchase.

2) *Keep up with trends in buyer preferences.* Constant monitoring of customer characteristics and purchases is crucial. Advertising and promotion never end.

Chain stores now carry nursery items. Convenience and escalating gas prices promote one-stop shopping.

3) *Know what combination of plants will maximise profits.* Ornamental plants fall into general categories of shade trees, conifers, perennials, vines, shrubs, bulbs and annuals. While most nurseries grow a range of plants, there is a trend toward specialisation. Growing only native ground covers or only daylilies are viable niche markets. The production of specialty crops, such as hardy bamboo and disease-free apple stock, and specialisation in plants in short supply, like uncommon plants and very large trees, are niche markets even small growers can serve.

Keep abreast of recent developments in the industry. Subscribing to trade publications and attending trade shows or conferences are good ways to learn about grower issues like plant availability, new varieties for specific needs, popular sizes, specific growing conditions and enhanced services. Enhanced services can include providing photographs of plants and making presentations to landscape architects and other potential customers.

Government agencies are increasingly concerned about invasive plant imports. Drug enforcement personnel are also concerned about the use of some plants. A concerted effort at the federal level to limit introductions of plant species to the United States raised concerns on the part of some plant explorers and nursery owners. However, it encouraged renewed interest in native species formerly put at a disadvantage by foreign imports, not only for in situ conservation, but also for increased use in landscaping. Several plants on invasive lists throughout the United States are important to the nursery trade. Plants such as Norway maple, butterfly bush, Japanese barberry and miscanthus have been lucrative sellers.

REFERENCES

Bell, C., M. Evans, and C. Wilen. 2000. "Exotic pest plants". *The Nursery Pro.* 2(2): 9, 17.

Derr. J.F., et.al.. 1997. "Weed management in landscape and nursery plantings" *in*: *Weed Management and Horticultural Crops*. ASHS Press. Pp. 42-59.

Rust, M.K., L. Greenberg, K. Kabashima, H. Costa, and C. Wilen. 2003. *Pest Management Alliance for the Containerized Nursery Industry.* Final Report to California Dept. Pesticide Regulation.

Sartain, A., C. A. Wilen, T. Ellis, R. Miller, and T. Salmon. 2006. "Tracking cottontail rabbit damage in a southern Californian nursery using GIS and GPS technology". Abst. 22nd Vertebrate Pest Conf. Berkeley, CA p. 31.

Stanton, A. E., C.E. Bell, and C. Wilen. 2002. *Keep it in the Garden: Invasive Plants and the Nursery Trade.* Proceedings 54th Cal. Weed Sci. Soc. 4p.

Wilen, C.A. 2003. IPM and water quality for nurseries. 2002 *UC IPM Update* p. 3.

11

Weed Control

A weed in a general sense is a plant that is considered by the user of the term to be a nuisance, and normally applied to unwanted plants in human-made settings such as gardens, lawns or agricultural areas, but also in parks, woods and other natural areas. More specifically, the term is often used to describe native or nonnative plants that grow and reproduce aggressively.Generally, a weed is a plant in an undesired place.

Weeds may be unwanted for a number of reasons: they might be unsightly, or crowd out or restrict light to more desirable plants or use limited nutrients from the soil. They can harbor and spread plant pathogens that infect and degrade the quality of crop or horticultural plants. Some weeds are a nuisance because they have thorns or prickles, some have chemicals that cause skin irritation or are hazardous if eaten, or have parts that come off and attach to fur or clothes.

The term weed in its general sense is a subjective one, without any classification value, since a "weed" is not a weed when growing where it belongs or is wanted. Indeed, a number of "weeds" have been used in gardens or other cultivated-plant settings. An example is the corncockle, Agrostemma, which was a common field weed exported from Europe along with wheat, but now sometimes grown as a garden plant.

Farmers would generally agree that weeds are not in the field because of a deficiency of herbicides or cultivation. Rather, weeds are the natural result of defying nature's preference for high species diversity and covered ground. Nature is trying to move the system in one direction, the farmer in another. We create weed problems through conventional crop production methods. After we create these problems, we spend huge sums of money and labour trying to "control" them. The opposite of reactive thinking is proactive thinking, by which we seek what we want through effective design and planning.

Weedy plants generally share similar adaptations that give them advantages and allow them to proliferate in disturbed environments whose soil or natural vegetative cover has been damaged. Naturally occurring disturbed environments include dunes and other windswept areas with shifting soils, alluvial flood plains, river banks and deltas, and areas that are often burned. Since human agricultural practices often mimic these natural environments where weedy species have evolved, weeds have adapted to grow and proliferate in human-disturbed areas such as agricultural fields, lawns, roadsides, and construction sites.

The weedy nature of these species often gives them an advantage over more desirable crop species because they often grow quickly and reproduce quickly, have seeds that persist in the soil seed bank for many years, or have short lifespans with multiple generations in the same growing season. Perennial weeds often have underground stems that spread out under the soil surface or, like ground ivy (Glechoma hederacea), have creeping stems that root and spread out over the ground. A number of weedy species have developed allelopathy, chemical means to prevent the germination or growth of neighboring plants.

As long as humans have cultivated plants, weeds have been a problem. Weed seeds are often collected and transported with crops after the harvesting of grains, and so many weed species have moved out of their natural geographic locations and have spread around the world with humans. Not all weeds have the same ability to damage crops and horticultural plants. Some have been classified as noxious weeds because if left unchecked, they often dominate the environment where crop plants are to be grown.

They are often foreign species mistakenly or accidentally imported into a region where there are few natural controls to limit their spread and population. Many weeds have ideal areas for growth and reproduction thanks to large areas of open soil created by the conversion of land to agriculture, and human distribution of food crops mixed with seeds of weeds from other parts of the world. Thus humans are the vector of transport and the producer of disturbed environments, so weedy species have an ideal association with humans.

A number of weeds, such as the dandelion Taraxacum, are edible, and their leaves and roots may be used for food or herbal medicine. Burdock is common weed over much of the world, and is sometimes used to make soup and other medicine in East Asia. These so-called "beneficial weeds" may have other beneficial effects, such as drawing away the attacks of crop-destroying insects, but often are breeding grounds for insects and pathogens that attack other plants. Dandelions are one of several species which break up hardpan in overly cultivated fields, helping crops grow deeper root systems. Some modern species of domesticated flower actually originated as weeds in cultivated fields and have been bred by people into garden plants for their flowers or foliage.

TYPES OF WEEDS

Weeds can be divided into two broad categories —annuals and perennials. Annual weeds are plants that produce a seed crop in one year, then die. They are well adapted to succeed in highly unstable and unpredictable environments brought about by frequent tillage, drought, or other disturbance. They put much of their life cycle into making seed for the next generation. This survival strategy serves plants in disturbed environments well, since their environment is likely to be disturbed again. The annual plant must make a crop of seed as soon as possible before the next disturbance comes. Annual plants also yield more seed than do perennial plants, which is why humans prefer annual over perennial crops for grain production. When we establish annual crop plants using tillage we also create an environment desirable for annual weeds.

Perennial weeds prosper in less-disturbed and more stable environments. They are more common under no-till cropping systems. Their objective is to put some energy into preserving the parent plant while producing a modest amount of seed for future generations. After a field is converted from conventional tillage to no-till, the weed population generally shifts from annual to perennial weeds.

Perennial weeds possess many of the characteristics of annual weeds: competitiveness, seed dormancy, and long-lived seed. In addition to these characteristics, many perennial weeds possess perennating parts such as stolons, bulbs, tubers, and rhizomes. These parts allow the parent plant to regenerate if damaged and to produce new plants from the parent plant without seed. Additionally, the perennating parts serve as food storage units that also enhance survival. These stored-food reserves allow for the rapid regrowth perennial weeds are known for.

CAUSE OF WEEDS

When a piece of land is left fallow, it is soon covered over by annual weeds. If the field is left undisturbed for a second year, briars and brush start to grow. As the fallow period continues, the weed community shifts increasingly toward perennial vegetation. By the fifth year, the field will host large numbers of young trees in a forest region, or perennial grasses in a prairie region. This natural progression of different plant and animal species over time is a cycle known as *succession*. This weed invasion, in all its stages, can be viewed as nature's means of restoring stability by protecting bare soils and increasing biodiversity.

Weeds are evidence of nature struggling to bring about ecological succession. When we clear native vegetation and establish annual crops, we are holding back natural plant succession, at great cost in weed control. To better understand this process, think of succession as a coil spring.

Managing cropland as an annual monoculture compresses the spring—leaving it straining to release its energy as a groundcover of weeds. In contrast, a biodiverse perennial grassland or forest is like the coil spring in its uncompressed condition—a state of relative stability with little energy for drastic change.

Generally speaking, biodiversity leads to more stability for the ecosystem as a whole. Modern crop agriculture is typified by large acreages of a single plant type, accompanied by a high percentage of bare ground—the ideal environment for annual weeds to prosper in the Weed seed distribution and density in agricultural soils are influenced by cropping history and the management of adjacent landscapes, and may be highly variable.

Western Nebraska cropland found 140 seeds per pound of surface soil, equivalent to 200 million seeds per acre. Redroot pigweed and common lambsquarter accounted for 86%. Growing without competition from other plants, a single redroot pigweed plant can produce more than 100,000 seeds, while a common lambsquarter plant can produce more than 70,000 seeds. New weed species can enter fields by many routes. Equipment moved from one field to the next—especially harvest equipment—spreads weed seeds, as does hay brought from one farm to another.

Crop seed is often contaminated with weed seed, and livestock transport weed seeds from one farm to another in their digestive tracts and in their hair. Practical actions that can be taken to prevent the introduction and spread of weeds include the use of clean seed , cleaning equipment before moving from one field to the next, and composting manures that contain weed seeds before applying them to the field. Survival and germination of weed seeds in the soil depend on the weed species, depth of seed burial, soil type, and tillage. Seeds at or near the soil surface can easily be eaten by insects, rodents, or birds. Also, they may rot or germinate.

Buried seeds are more protected from seedeating animals and buffered from extremes of temperature and moisture. On average, about 4% of broadleaf and 9% of grass weed seeds present in the soil germinate in a given year. Results from seed burial experiments demonstrated that seeds of barnyard grass and green foxtail buried at 10 inches showed germination rates of 34 to 38% when dug up and spread on the soil surface. In the same study, seed buried at one inch showed only one to five percent germination.

Seeds were buried at different depths for a period of three years. Seed germination was greater with increasing depth of burial. These studies show that seeds near the surface face lots of hazards to their survival, while those buried deeply by tillage are more protected. When those deep-buried seed are plowed up to the surface again they have a good chance of germinating and growing.

A higher percentage of seed remains near the soil surface under chisel plow and no-till. The moldboard plow and ridge-till systems are stirring the soil more, burying lots of weed seeds, and keeping weed seed more evenly distributed down to a six-inch depth. After a seed is shed from the parent plant, it can remain dormant or germinate. There are several different types of dormancy. Seeds with hard seed coats possess "innate" dormancy.

Several weed species, including pigweed, have seed coats that require mechanical or chemical injury and high-temperature drying to break dormancy. Another type of innate dormancy can best be described as after-ripening, meaning the seed requires further development after it falls off the plant before it will germinate. Several grass and mustard family weeds require after-ripening. "Induced" dormancy results when seeds are exposed to unfavourable conditions, such as high temperatures, after being shed from the parent plant. "Enforced" dormancy occurs when conditions favorable to weed germination are absent.

The seeds remain dormant until favorable conditions return. Altogether, multiple types of dormancy ensure that some weed seeds will germinate and some will remain dormant for later seasons. Some weed species are dependent on light for germination; some germinate in either light or darkness; others germinate only in the dark. Thus, there are no hard and fast rules for managing an overall weed population according to light sensitivity. Manure application may stimulate weed germination and growth. Studies have shown that poultry manure does not contain viable weed seeds, yet weed levels often increase rapidly in pastures following poultry manure application. Since chickens and turkeys have a gizzard capable of grinding seeds, weed seeds are not likely to pass through their digestive systems intact.

Additionally, most poultry rations contain few if any weed seeds. The weed germination is probably caused by effects of ammonia on the weed-seed bank already present in the soil. The effect varies depending on the source of the litter and the weed species present. Manure from hoofed livestock (e.g., sheep, cattle, and horses), on the other hand, may indeed contain weed seed that has passed through their digestive systems.

Composted manure contains far fewer weed seeds than does raw manure because the heat generated during thc composting process kills them. Fertilisation practices can also affect weed germination. Where fertiliser is broadcast, the entire weed community is fertilised along with the crop. Where fertiliser is banded in the row, only the crop gets fertilised.

EFFECTS OF WEEDS ON OTHER PLANTS

Weeds can compete with productive crops or pasture, or convert productive land into unusable scrub. Weed are also often poisonous, distasteful, produce burrs, thorns or other damaging body parts or otherwise interfere with the use and management of desirable plants by contaminating harvests or excluding livestock.

Weeds tend to thrive at the expense of the more refined edible or ornamental crops. They provide competition for space, nutrients, water and light, although how seriously they will affect a crop depends on a number of factors. Some crops have greater resistance than others- smaller, slower growing seedlings are more likely to be overwhelmed than those that are larger and more vigorous. Onions are one of the crops most susceptible to competition, for they are slow to germinate and produce slender, upright stems. Quick growing, broad leafed weeds therefore have a distinct advantage, and if not removed, the crop is likely to be lost. Broad beans however produce large seedlings, and will suffer far less profound effects of weed competition other than during periods of water shortage at the crucial time when the pods are filling out. Transplanted crops raised in sterile seed or potting compost will have a head start over germinating weed seeds.

Weeds also differ in their competitive abilities, and can vary according to conditions and the time of year. Tall growing vigorous weeds such as fat hen (Chenopodium album) can have the most pronounced effects on adjacent crops, although seedlings of fat hen that appear in late summer will only produce small plants. Chickweed (Stellaria media), a low growing plant, can happily co-exist with a tall crop during the summer, but plants that have overwintered will grow rapidly in early spring and may swamp crops such as onions or spring greens.

The presence of weeds does not necessarily mean that they are competing with a crop, especially during the early stages of growth when each plant can find the resources it requires without interfering with the others. However, as the seedlings' size increases, their root systems will spread as they each begin to require greater amounts of water and nutrients. Estimates suggest that weed and crop can co-exist harmoniously for around three weeks, therefore it is important that weeds are removed early on in order to prevent competition occurring. Weed competition can have quite dramatic effects on crop growth. Harold A Roberts cites research carried out with onions wherein "Weeds were carefully removed from separate plots at different times during the growth of the crop and the plots were then kept clean. It was found that after competition had started, the final yield of bulbs was being reduced at a rate equivalent to almost 4% per day. So that by delaying weeding for another fortnight, the yield was cut to less than half that produced on ground kept clean all the time." He goes on to record that "by early June, the weight of weeds per unit area was twenty times that of the crop, and the weeds had already taken from the soil about half of the nitrogen and a third of the potash which had been applied".

Perennial weeds with bulbils, such as lesser celandine and oxalis, or with persistent underground stems such as couch grass (Agropyron repens) or creeping buttercup (Ranunculus repens) are able to store reserves of food, and are thus able to grow faster and with more vigour than their annual counterparts. There is also evidence that the roots of some perennials such as couch grass exude allelopathic chemicals which inhibit the growth of other nearby plants.

Weeds can also host pests and diseases that can spread to cultivated crops. Charlock and Shepherd's purse may carry clubroot, eelworm can be harboured by chickweed, fat hen and shepherd's purse, while the cucumber mosaic virus, which can devastate the cucurbit family, is carried by a range of different weeds including chickweed and groundsel.

However, at times the role of weeds in this respect can be over-rated. As far as insect pests are concerned, often the species that live on weeds are not the same as those that attack vegetable crops; "Tests with the common cruciferous weeds such as shepherds purse have shown that they do not act as hosts for the larvae of the cabbage root fly. One exception was found to be the wild radish, but this is not usually a weed of established vegetable gardens". However pests such as cutworms may first attack weeds then move on to cultivated crops.

While charlock, a common weed in southeastern USA, may be considered a weed by row crop growers, it is highly valued by beekeepers, who seek out places where it blooms all winter, thus providing pollen for honeybees and other pollinators. Its bloom is resistant to all but a very hard freeze, and even that will only kill it back briefly. By feeding an array of pollinators during a seasonal dearth, it can redound to the farmer's advantage. Many weeds are likewise highly beneficial to pollinators.

PROACTIVE WEED MANAGEMENT STRATEGIES

In domestic gardens, methods of weed control include covering an area of ground with several layers of wet newspaper or one black plastic sheet for several weeks. In the case of using wet newspaper, the multiple layers prevent light from reaching all plants beneath, which kills them. Saturating the newspaper with water daily speeds the decomposition of the dead plants. Any weed seeds that start to sprout because of the water will also be deprived of sunlight, be killed, and decompose. After several weeks, all germinating weed seeds present in the ground should be dead. Then the newspaper can be removed and the ground can be planted. The decomposed plants will help fertilise the plants or seeds planted later.

In the case of using the black plastic sheet, the greenhouse effect is used to kill the plants beneath the sheet. A 5-10 cm layer of wood chip mulch on the ground will also prevent most weeds from sprouting. Also, gravel can be spread over the ground as an inorganic mulch. In agriculture, irrigation is sometimes used as a weed control measure such as in the case of paddy fields. Many people find that although the black plastic sheeting is extremely effective at preventing the weeds in areas where it covers, in actual use it is difficult to achieve full coverage.

Knowing how weeds reproduce, spread and survive adverse conditions can help in developing effective control and management strategies. Weeds have a range of techniques that enable them to thrive;

Annual and biennial weeds such as chickweed, annual meadow grass, shepherd's purse, groundsel, fat hen, cleaver, speedwell and hairy bittercress propagate themselves by seeding. Many produce huge numbers of seed several times a season, some all year round. Groundsel can produce 1000 seed, and can continue right through a mild winter, whilst scentless mayweed produces over 30,000 seeds per plant. Not all of these will germinate at once, but over several seasons, lying dormant in the soil sometimes for years until exposed to light. Poppy seed can survive 80-100 years, dock 50 or more. There can be many thousands of seeds in a square foot or square metre of ground, thus and soil disturbance will produce a flush of fresh weed seedlings.

One technique employed by growers is the 'stale seed bed', which involves cultivating the soil, then leaving it for a week or so.When the initial flush of weeds has germinated, the grower will lightly hoe off before the desired crop is planted. However, even a freshly cleared bed will be susceptible to airborne seed from elsewhere, as well as seed brought in by passing animals which can carry them on their fur, or from freshly imported manure. The organic solution to the problem of spreading annual weeds lies in regular, properly timed weeding, preferably just before flowering (fortuitously, this is also the time at which they will be of the most value in composting).This technique is also quite often used by farmers who let weeds germinate then return the soil before crop sowing.

Perennial weeds also propagate by seeding; the airborne seed of the dandelion and the rose-bay willow herb are parachuted far and wide. But they also have an additional range of vegetative means of spreading that gives them their pernicious reputation. Dandelion and dock put down deep tap roots, which, although they do not spread underground, are able to regrow from any remaining piece left in the ground. Removal of the complete tap root is the only sure remedy.

The most persistent of the perennials are those that spread by underground creeping rhizomes that can regrow from the tiniest fragment. These include couch grass, bindweed, ground elder, nettles, rosebay willow herb, Japanese knotweed, horsetail and bracken, as well as creeping thistle, whose tap roots can put out lateral roots. Other perennials put out runners that spread along the soil surface. As they creep along they set down roots, enabling them to colonise bare ground with great rapidity. These include creeping buttercup and ground ivy. Yet another group of perennials propagate by stolons- stems that arch back into the ground to reroot. Most familiar of these is the bramble.

All of the above weeds can be very difficult to eradicate- thick black plastic mulches can be effective to a degree, although will probably need to be left in place for at least two seasons. In addition, hoeing off weed leaves and stems as soon as they appear can eventually weaken and kill the plants, although this will require persistence in the case of plants such as bindweed. Nettle infestations can be tackled by cutting back at least three times a year, repeated over a three year period. Bramble can be dealt with in a similar way. Some plants are said to produce root exudates that suppress herbaceous weeds. Tagetes minuata is claimed to be effective against couch and ground elder, whilst a border of comfrey is also said to act as a barrier against the invasion of some weeds including couch.

In the preceding sections, we saw how weeds are established and maintained by human activities. So, how do we begin to manage an unnatural system to our best benefit without compromising the soil and water? We can start by putting the principles of ecology to work on our behalf, while minimising actions that only address symptoms.

Crops that kill Weeds

Some crops are especially useful because they have the ability to suppress other plants that attempt to grow around them. Allelopathy refers to a plant's ability to chemically inhibit the growth of other plants. Rye is one of the most useful allelopathic cover crops because it is winter-hardy and can be grown almost anywhere.

Rye residue contains generous amounts of allelopathic chemicals. When left undisturbed on the soil surface, these chemicals leach out and prevent germination of small-seeded weeds. Weed suppression is effective for about 30 to 60 days. If the rye is tilled into the soil, the effect is lost. A weed scientist in Michigan observed that some large-seeded food crops planted into rye mulch had high tolerance to the allelopathic effects, while smaller-seeded crops had less tolerance.

In the corn, cucumber, pea, and snapbean no-till planted under rye mulch germinated and grew as well or better than the same crops planted no-till without mulch. Smaller-seeded crops, including cabbage and lettuce, showed much less germination, growth, and yield. Weeds that were reduced by rye mulch included ragweed (by 43%), pigweed (95%), and common purslane (100%). Dr. Doug Worsham, a North Carolina weed scientist, concluded that leaving a small grain mulch and not tilling gives 75 to 80% early-season reduction of broadleaf weeds. North Carolina researchers investigated combinations of herbicide use and cover crop plantings on weed control. Rye and subterranean clover showed the highest weed control without herbicides. Neither provided as much control as herbicides, however.

Tillage reduced weed control considerably· where no herbicide was used, as compared to no-tillage. By season's end the weed control resulting from cover crops alone had decreased. The researchers concluded that additional weed control measures must be applied with cover crops to assure effective weed control and profitable yields. Other crops that have shown allelopathic effects include sunflowers, sorghum, and rapeseed.

Weed control ability varies among varieties and management practices. Sweet potatoes have been shown to inhibit the growth of yellow nut-sedge, velvetleaf, and pigweed. Field trials showed a 90% reduction of yellow nutsedge over two years following sweet potatoes. Rapeseed, a type of mustard, has been used to control weeds in potatoes and corn under experimental conditions. All members of the mustard family (Brassicaceae) contain mustard oils that inhibit plant growth and seed germination. The concentration of allelopathic mustard oils varies with species and variety of mustard.

Researchers have begun to study ways to manage mustard's weed-suppressive abilities in crop production. In a Pacific Northwest study, fallplanted 'Jupiter' rapeseed and sundangrass were evaluated for suppression of weeds growing in spring-planted potatoes. In the spring, the researchers either tilled or strip-killed the rapeseed in preparation for potato planting. The first year of the study, rapeseed reduced midseason weed production 85% more than fallowing. By the end of the season, weed production was reduced by 98% with rapeseed, but only 50% the second year. In general, typical levels of cover crop residues, when left on the soil surface, can be expected to reduce weed emergence by 75 to 90%. As these residues decompose, the weed suppression effect will decline also. Residues that are more layered and more compressed will be more suppressive.

Small-seeded weeds that have light requirements for sprouting are most sensitive to cover crop residue. Larger-seeded annual and perennial weeds are least sensitive to residue. Effective management strategies include growing cover crops that produce high amounts of residue, growing slower-decomposing cover crops, packing the mulch down with implements that compress it, and using methods other than cover crops to control largeseeded annual and perennial weeds.

Smother Crops and Mulch

Certain crops can be used to smother weeds. Short-duration plantings of buckwheat and sorghum-sudangrass, for example, smother weeds by growing faster and out-competing them. In northern states, oats are commonly planted as a "nurse crop" for alfalfa, clover, and legumegrass mixtures—the oats simply take the place of weeds that would otherwise grow between the young alfalfa plants.

With enough mulch, weed numbers can be greatly reduced. Nebraska scientists applied wheat straw in early spring to a field where wheat had been harvested the previous August. At the higher straw rates, weed levels were reduced by more than two thirds. Wheat, like rye, is also known to possess allelopathic qualities, which may have contributed to the weed suppression.

Crop Rotations

Crop rotations limit the buildup of weed populations and prevent major weed species shifts. Weeds tend to prosper in crops that have requirements similar to the weeds. Fields of annual crops favour short-lived annual weeds, whereas maintaining land in perennial crops favours perennial weed species.

Two examples would be shattercane in continuous sorghum and downy brome in continuous winter wheat. In a crop rotation, the timing of cultivation, mowing, fertilisation, herbicide application, and harvesting changes from year to year. Rotation thus changes the growing conditions from year to year—a situation to which few weed species easily adapt. Rotations that include clean-cultivated annual crops, tightly spaced grain crops, and mowed or grazed perennial sod crops create an unstable environment for weeds.

Additional weed control may be obtained by including short-season weeds mothering crops such as sorghum-sudan or buckwheat. Crop rotation has long been recognised for this ability to prevent weeds from developing to serious levels. In a dryland wheat study, continuous winter wheat was compared to a rotation of winter wheat/proso millet/fallow or a winter wheat/ sunflower/fallow rotation. The year before, at the start of the study, the fields were in winter wheat and were sprayed with Roundup™ and 2,4-D. The sunflowers were treated with Prowl™. All other weed control was by mechanical means, including a sweep and rodweeder as needed.

During the two-year study, weed levels were 145 plants per square yard for the continuous wheat, 0.4 plants per square yard for the winter wheat/proso millet fallow system, and 0.3 for the winter wheat/ sunflower fallow system.

Intercropping

Intercropping can be used as an effective weed control strategy. Having different plant types growing together enhances weed control by increasing shade and increasing crop competition with weeds through tighter crop spacing. Where one crop is relayintercropped into another standing crop prior to harvest, the planted crop gets off to a weed-free start, having benefited from the standing crop's shading and competition against weeds. Such is the case when soybeans are interplanted into standing green wheat—the thick wheat stand competes well with weeds while the soybeans are getting started.

Planting method, planting date, and variety must be well-planned in advance. Though soybeans can be directly drilled into the standing green wheat, less wheat damage occurs if the wheat is planted in skiprows. Skiprows are created by plugging certain drop tube holes in the grain drill. Soybeans can be planted with row units set at spacings matched to the skiprows in the wheat. If the wheat is drilled on 7½-inch rows, to create a 30-inch row spacing for soybeans, every fourth drill hole in the wheat drill would be plugged. Tractor tires will follow the skips, resulting in no damage to the wheat. Studies in Missouri and Ohio showed that wheat yields were three to six bushels per acre less when intercropped with soybeans than when solid-drilled and grown alone. Generally, soybean yields are higher when intercropped into wheat than when double cropped behind wheat in the central and northern Midwest, where double cropping is risky due to a shorter growing season.

Weed-free by Design

Thus far, we've seen that weeds are a symptom of land management that defies nature's design. Stirring the soil with tillage creates conditions favorable for weed germination and survival. Monocultures of annual crops hold natural plant succession back and minimise biodiversity, inviting weed populations to thrive. When we try to maintain bare ground, weeds grow to cover the soil and increase biodiversity. If we take a proactive approach to the whole agricultural system, rather than just looking at the parts, we can use the principles of nature to our advantage instead of fighting them. We will never win the war against nature, and, she has much more patience than we do. When we try to break the rules of nature, we end up breaking ourselves against the rules. Let's look at an agronomic system where—by design—weeds simply are not a problem.

One of the biggest shortcomings in American agriculture is the separation of plant and animal production. Commodity crop production of corn, milo, and soybeans is really a component of animal production because these crops are largely fed to livestock. It seems inefficient to grow grains separately and haul them to animal-feeding facilities. The milo and cowpeas are so vigorous they outcompete any weeds present. Here nature's principle of biodiversity is obeyed rather than fought with herbicides.

Previously, when the milo was grown separately, he had to spray for greenbugs. After he started with the pea-milo mixture, the greenbug problem disappeared. When the milo and peas are mature, he combines them. This produces a milo to pea ratio of 2/3 to 1/3, which is ideal for feed. After grain harvest he turns his animal mixture of hogs, cattle, sheep, and chickens into the standing crop stubble, thereby adding more diversity. All the waste grain is consumed by livestock, and the stubble trampled into the soil, at a profit in animal gains to the farmer. What few weeds may have grown up with the crop can be eaten by the livestock. Under typical single-crop scenarios, the waste grain would rot in the field and the farmer might incur a $6/acre stalk mowing cost. In this case, following the principle of biodiversity increased profit by lowering cost. Bill and his team designed weeds out of the system. Other opportunities exist to design weeds out of the farming operation. These opportunities are limited only by human creativity—the most underutilised tool in the toolbox.

Reactive Measures

The reactive paradigm of weed management is typified by the word *control*. This word assumes that weeds are already present, or to be expected, and the task is to solve the problem through intervention. Agriculture magazines are chock-full of advertisements promising season-long control, complete control, and control of your toughest weeds. These ads imply that the secret is in the proper tank mix of herbicides.

Examining these ads from a cause-andeffect standpoint, we might well conclude that weeds are caused by a deficiency of herbicides in the field. When selecting a tool for weed management, it helps to understand the weed's growth stages and to attack its weakest growth stage.

Alternatively, management techniques that discourage weed seed germination could be implemented. In so doing, a farmer can identify a means of control that requires the least amount of resources. The various tools available for weed management fall into two categories: those that enhance biodiversity in the field and those that reduce it. This is not to imply a "good vs. bad" distinction. Rather it is meant to describe the effect of the tool on this important characteristic of the crop/weed interaction. In general, as plant diversity increases, weeds become less of a problem.

WEED CONTROL TOOLS

Herbicides

Since herbicide information is abundantly available from other sources, it is not covered in detail in this guide. Herbicides can be effective in maintaining ground cover in no-till systems by replacing tillage operations that would otherwise create bare ground and stimulate more weed growth. Until better weed management approaches can be found, herbicides will continue to remain in the toolbox of annual crop production. However, some farmers are realising that with continued herbicide use, the weed problems just get worse or at best stay about the same. Nature never gives up trying to fill the vacuum created by a simplified bare-ground monoculture, and long-term use of the same herbicide leads to resistant weeds, as they adapt to the selection pressure applied to them. But compared to tillage systems where bare ground is maintained, herbicide use may be considered the lesser of two evils. At least where ground cover is maintained, the soil is protected from erosion for future generations to farm.

Herbicides are generally classified as follows:

- Contact herbicides destroy only that plant tissue in contact with the chemical spray. Generally, these are the fastest acting herbicides. They are ineffective on perennial plants that are able to re-grow from roots or tubers.
- Systemic herbicides are foliar-applied and are translocated through the plant and destroy a greater amount of the plant tissue. Modern herbicides such as · glyphosate are designed to leave no harmful residue in the soil.
- Soil-borne herbicides are applied to the soil and are taken up by the roots of the target plant.
- Pre-emergent herbicides are applied to the soil and prevent germination or early growth of weed seeds.

In agriculture large scale and systematic weeding is usually required, often by machines, such as liquid herbicide sprayers, or even by helicopter, to eliminate the massive amount of weeds present on farming lands.

However there are a number of techniques that the organic farmer can employ such as mulching and carefully timed cutting of weeds before they are able to set seed.

Least-toxic Herbicides

Corn gluten meal has been used successfully on lawns and high-value crops as a pre-emergent herbicide. It must be applied just prior to weed seed germination to be effective. A common rate is 40 pounds per 1000 square feet, which suppresses many common grasses and herbaceous weeds. Two name brand weed control products containing corn gluten meal are WeedBan™ and Corn Weed Blocker™. Herbicidal soaps are available from Ringer Corporation and from Mycogen. Scythe™, produced by Mycogen, is made from fatty acids. Scythe acts fast as a broad-spectrum herbicide, and results can often be seen in as little as five minutes. It is used as a post-emergent, sprayed directly on the foliage.

Vinegar is an ingredient in several new herbicides on the market today. Burnout™ and Bioganic™ are two available brands. Both of these are post-emergent burndown herbicides. They are sprayed onto the plant to burn off top growth—hence the concept "burndown." As for any root-killing activity with these two herbicides, I cannot say. The label on Burnout™ states that perennials may regenerate after a single application and require additional treatment.

Researchers in Maryland tested 5% and 10% acidity vinegar for effectiveness in weed control. They found that older plants required a higher concentration of vinegar to kill them. At the higher concentration, they got an 85 to 100% kill rate. A 5% solution burned off the top growth with 100% success. Household vinegar is about 5% acetic acid. Burnout™ is 23% acetic acid. Bioganic™ contains 10% acetic acid plus clove oil, thyme oil, and sodium lauryl sulfate.

AllDown Green Chemistry herbicide™ contains acetic acid, citric acid, garlic, and yucca extract. MATRAN™ contains 67% acetic acid and 34% clove oil. Weed Bye Bye™ contains both vinegar and lemon juice. Vinegar is corrosive to metal sprayer parts the higher the acidity, the more corrosive. Plastic equipment is recommended for applying vinegar. Dr. Jorge Vivanco of Colorado State University horticulture department isolated the compound "catechin," a root exudate from spotted knapweed, *Centaurea maculosa,* that has strong herbicidal properties. Knapweed uses the compound as an allelopathic method of competing with other plants. Several companies are interested in producing an environmentally friendly natural herbicide from the root exudate. Since catechin is naturally occuring, new herbicides made from it may be eligible for EPA's fast-track approval process.

Weeder Geese

Weeder geese have been used successfully both historically and in more recent times. They are particularly useful on grass weeds in a variety of crops. Chinese or African geese are favorite varieties for weeding purposes. Young geese are usually placed in the fields at six to eight weeks of age. They work well at removing weeds between plants in rows that cannot be reached by cultivators or hoes. If there are no trees in the field, temporary shade will be needed.

Supplemental feed and water must be provided as well. Water and feed containers can be moved to concentrate the geese in a certain area. A 24- to 30-inch fence is adequate to contain geese. Marauding dogs and coyotes can be a problem and should be planned for with electric fencing or guard animals. At the end of the season, bring geese in for fattening on grain.

Carrying geese over to the next season is not recommended, because older geese are less active in hot weather than younger birds. Additionally, the cost of overwintering them outweighs their worth the next season. Geese have been used on the following crops: cotton, strawberries, tree nurseries, corn, fruit orchards, tobacco, potatoes, onions, sugar beets, brambles, other small fruits, and ornamentals.

Tillage

Tillage and cultivation are the most traditional means of weed management in agriculture. Both expose bare ground, which is an invitation for weeds to grow. Bare ground also encourages soil erosion, speeds organic matter decomposition, disturbs soil biology, increases water runoff, decreases water infiltration, damages soil structure, and costs money to maintain. Some specific tillage guidelines and techniques for weed management include the following:

- *Preplant tillage.* Where weeds such as quackgrass or johnsongrass exist, spring-tooth harrows and similar tools can be effective in catching and pulling the rhizomes to the soil surface, where they desiccate and die. Discing, by contrast, tends to cut and distribute rhizomes and may make the stand even more dense.
- *Blind tillage.* Blind cultivation is a pre-emergent and early post-emergent tillage operation for weed control. It usually employs either finger weeders, tine harrows, or rotary hoes. These implements are run across the entire field, including directly over the rows. The large-seeded corn, soybeans, or sunflowers survive with minimal damage, while small-seeded weeds are easily uprooted and killed. For corn, the first pass should be made between three and five days after planting and a second at the spike stage. Blind cultivation may continue until the crop is about five inches tall. For soybeans, the first pass should be done when germinating crop seedlings are still about ½ inch below the soil surface, but not when the "hook" is actually emerging. The second pass should be done three to five days after soybean emergence, and twice later at four-day intervals. Sunflowers can be blind-tilled up to the six-leaf stage, giving them an excellent head start on weeds. Grain sorghum may be rotary hoed prior to the spike stage, and again about one week after spike stage. Because the seed is small, timing for blind-till in sorghum is very exacting. Post-emergent blind tillage should be done in the hottest part of the day, when crop plants are limber, to avoid excessive damage. Rotary hoes, not harrows, should be used if the soil is crusted or too trashy. Seeding rates should be increased 5 to 10% to compensate for losses in blind cultivation.
- *Row crop cultivation.* Cultivation is best kept as shallow as possible to bring as few weed seeds as possible to the soil surface. Where perennial rhizome weeds are a problem, the shovels farthest from the crop row may be set deeper on the first cultivation to bring rhizomes to the surface. Tines are more effective than duck feet sweeps for this purpose. Later cultivations should have all shovels set shallow to avoid excessive pruning of crop roots. Earliest cultivation should avoid throwing soil toward the crop row, as this places new weed seed into the crop row where it may germinate before the crop canopy can shade it out. Use row shields as

appropriate. As the crop canopy develops, soil should be thrown into the crop row to cover emerging weeds.

– *Interrow cultivation* is best done as soon as possible after precipitation, once the soil is dry enough to work. This avoids compaction, breaks surface crusting, and catches weeds as they are germinating—the most vulnerable stage.

Generally speaking, tillage systems tend to discourage most biennial and perennial weed species, leaving annual weeds as the primary problem. Exceptions to this are several weeds with especially resilient underground rhizome structures such as johnsongrass, field bindweed, and quackgrass. Plowing of fields to bring up the rhizomes and roots has been used to control bindweed and quackgrass.

Another interesting application of timing to weed control is night tillage. Researchers have found that germination of some weed species is apparently triggered by exposure to light. Tillage done in darkness exposes far fewer seeds to light and reduces weed pressure. So far, small seeded broadleaf weeds (lambsquarter, ragweed, pigweed, smartweed, mustard, and black nightshade) appear to be most readily affected.

Flame Weeding

Preplant, pre-emergent, and post-emergent flame weeding has been successful in a number of crops. The preplant application has commonly been referred to as the "stale seedbed technique." After seedbed tillage is completed, weed seeds, mostly in the upper two inches of the soil, are allowed to sprout. Assuming adequate moisture and a minimum soil temperature of 50º F, this should occur within two weeks.

A fine to slightly compacted seedbed will germinate a much larger number of weeds. The weeds are then "seared" with a flamer, or burned down with a broadspectrum herbicide, preferably when the populationInsteIn, Instead, the ridge-till planter population is between the first and fifth trueleaf stages, a time when they are most susceptible. The crop should then be seeded as soon as possible, and with minimal soil disturbance to avoid bringing new seed to the surface. For the same reason, subsequent cultivations should be shallow. Pre-emergent flaming may be done after seeding, and in some crops post-emergent flaming may be done as well.

Flaming is often used as a band treatment for the crop row, and usually combined with interrow cultivation. Early flaming may be done in corn when it is 1.5 to 2 inches high. The growing tip is beneath the soil surface at this stage, and the crop readily recovers from the leaf damage. Subsequent post-emergent flamings may be done when corn reaches 6–10 inches in height, and later at layby. No flaming should be done when corn is at approximately 4 inches high, as it is most vulnerable then. The burners are offset to reduce turbulence and to avoid concentrating too much heat on the corn. Water shields are available on some flame weeder models. Uniform seedbed preparation and uniform tractor speed are important elements in flaming. Hot and dry weather appears to increase the efficacy of flaming.

Searing the plant is much more successful than charring. Excessive burning of the weeds often stimulates the roots and encourages regrowth, in addition to using more fuel. Flaming has generally proved most successful on young broadleaf weeds. It is reportedly less successful on grasses, as the seedlings develop a protective sheath around the growing tip when they are about 1 inch tall. Some concerns with the use of fire include possible crop damage, potential dangers in fuel handling, and the cost of fuel.

Organic Methods

Typically a combination of methods are used in organic situations.

- *Drip irrigation*: Rubber hoses and other methods are used to bring water directly to the roots of the desired plants. This limits weed access to water.
- *Manually pulling weeds*: Labourers are used to pull weeds at various points in the growing process.
- *Mechanically tilling around plants*: Tractors are used to carefully till weeds around the crop plants at various points in the growing process. Besides tilling, other mechanical weed control methods also exist
- *Ploughing*: Ploughing includes tilling of soil, intercultural ploughing and summer ploughing. Ploughing through tilling of soil uproots the weeds which causes them to die. In summer ploughing is done during deep summers. Summer ploughing also helps in killing pests.
- *Crop rotation*: Rotating crops with ones that kill weeds by choking them out, such as hemp,Mucuna pruriens, and other crops, can be a very effective method of weed control. It is a way to avoid the use of herbicides, and to gain the benefits of crop rotation.
- *Weed mat*: A weed mat is an artificial mulch, fibrous cloth material, bark or newspaper laid on top of the soil preventing weeds from growing to the surface.

Woods

Besides those kinds of weeds which are of an herbaceous nature, there are others which are woody, and grow to a very considerable size; such as broom, furze and Lantanas. The first may be destroyed by frequent ploughing and harrowing, in the same manner as other perennial weeds are. Another method of destroying broom is by pasturing the field where it grows with grazers.

The best method of extirpating furze is to set fire to it in frosty weather, for frost has the effect of withering and making them burn readily. The stumps must then be cut over with a hatchet, and when the ground is well softened by rain it may be ploughed up, and the roots taken out by a harrow adapted to that purpose. If the field is soon laid down to grass, they will again spring up; in this case, pasturing with grazers is an effectual remedy. The thorn, or bramble, can only be extirpated by ploughing up the ground and collecting the roots.

In June, weeds are in their most succulent state, and in this condition, after they have lain a few hours to wither, cattle will eat almost every species. There is scarcely a hedge, border, or a nook, but what at that season is valuable.

INTEGRATED WEED MANAGEMENT

An integrated approach means assembling a weed management plan that incorporates a number of tools consistent with farm goals. Included are sanitation procedures, crop rotations, specialised tillage schemes, cover crops, and herbicides. The best examples of integrated approaches have been developed onfarm, by farmers themselves. Fields are overseeded or drilled in the fall with combinations of hairy vetch, oats, and grain rye as a winter cover crop.

The vetch provides nitrogen, while the grasses provide weed suppression and erosion protection. The cover crop is not tilled in before planting. skims off enough of the ridge top to create a clean seeding strip. Subsequent passes with the ridge-till cultivator eliminate any cover crop in the interrow area and help to re-shape the ridges. In Windsor, North Dakota, Fred Kirschenmann has developed a diverse rotation including cool-weather crops like oats, rye, barley, and spring wheat, and warm-season crops like sunflower, buckwheat, and millet. He employs selective timing to manage his principal weed problem, pigeon grass. By planting coolweather grains early, he can get a competitive jump on the weed, which requires somewhat warmer soil to germinate. The warm-season crops do best long after pigeon grass has germinated, however. He uses shallow pre-plant tillage to control weeds in these crops. One of the many advantages of composting is the reduction of viable weed seeds, which are killed by heat during the curing process. Don and Deloris Easdale of Hurdland, Missouri, reduced their annual herbicide costs from $10,000 to less than $1,000 in three years on their 300-plus acres of grain crops. They use hairy vetch, winter rye, or Austrian winter peas in combination with their ridge-till system.

They flail chop hairy vetch or winter peas ahead of the ridge-till planter and plant directly into the remaining cover crop residue. This practice eliminates using a burndown herbicide. The legumes replace much of the nitrogen needed for the corn or milo crop. Some liquid starter and liquid nitrogen is placed below the seed at planting.

They more than recover the seed costs of their cover crops in savings on fertiliser and herbicide.

REFERENCES

Elmore, C.L. and C. A. Wilen. 1997. *Weed management in landscapes*. UC IPM Pest Notes 7441. 5pp.

Masiunas, J. M. McGiffen, C. Wilen, C. Bell, T. Lanini, J. Derr, and G. Kolasani. 1997. "Integrated weed management in horticultural crops". *In*: *Weed Management and Horticultural Crops*. ASHS Press. Pp. 1-16.

Renz, M. C. Bell, C. Wilen, and J. DiTomaso. 2002. Perennial Pepperweed (Lepidium latifolium) Foreign invader in California's wildlands. *UC Weed Research Information Center Leaflet* #02-1.

Wilen, C.A. and C. L. Elmore. Revised in 2001. *Weed management in landscapes*. Pest Notes UC DANR Publication.

12

Pest Management

The goal of development is to maximise the use of energy, natural resources, capital and scientific information for the welfare of mankind. However, the process of developing agricultural production, water resource management, improvement of health and other activities of mankind, create an environment favorable to the development of organisms competing with man. This organism is designated as a pest but such a designation is not static, since a pest may be damaging and edible at the same time. Crickets and grasshoppers are acceptable as food by some people, but can be a curse to rice farmers. A pest problem exists when an organism interferes with human activities or desire, or otherwise competes with man. To rationally minimise or control pest depredations, an holistic approach to suppression is emphasized.

Pest control refers to the regulation or management of a species defined as a pest, usually because it is perceived to be detrimental to a person's health, the ecology or the economy. Pest control is at least as old as agriculture, as there has always been a need to keep crops free from pests. In order to maximize food production, it is advantageous to protect crops from competing species of plants, as well as from herbivores competing with humans.

The conventional approach was probably the first to be employed, since it is comparatively easy to destroy weeds by burning them or plowing them under, and to kill larger competing herbivores, such as crows and other birds eating seeds. Techniques such as crop rotation, companion planting (also known as intercropping or mixed cropping), and the selective breeding of pest-resistant cultivars have a long history.

Many pests have only become a problem because of the direct actions of humans. Modifying these actions can often substantially reduce the pest problem. In the USA, raccoons caused a nuisance by tearing open refuse sacks. Many householders introduced bins with locking lids, which deterred the raccoons from visiting. House flies tend to accumulate wherever there is human activity and is virtually a global phenomenon, especially where food or food waste is exposed. Similarly, seagulls have become pests at many seaside resorts. Tourists would often feed the birds with scraps of fish and chips, and before long, the birds would become dependent on this food source and act aggressively towards humans.In the UK, following concern about animal welfare, humane pest control and deterrence is gaining ground through the use of animal psychology rather than destruction. For instance, with the urban Red Fox which territorial behaviour is used against the animal, usually in conjunction with non-injurious chemical repellents.

Chemical pesticides date back 4,500 years, when the Sumerians used sulfur compounds as insecticides. The Rig Veda, which is about 4,000 years old, also mentions the use of poisonous plants for pest control. Ancient Chinese and Egyptian cultures are known to have used chemical pest controls[citation needed]. But it was only with the industrialization and mechanization of agriculture in the 18th and 19th century, and the introduction of the insecticides pyrethrum and derris that chemical pest control became widespread. In the 20th century, the discovery of several synthetic insecticides, such as DDT, and herbicides boosted this development. Chemical pest control is still the predominant type of pest control today, although its long-term effects led to a renewed interest in traditional and biological pest control towards the end of the 20th century.

TYPES OF PEST CONTROL

Organic Pest and Insect Control

While chemical pesticides may kill insects effectively, some may also be toxic to human beings and lead to severe environmental degradation if their use is not properly managed. By comparison, natural pesticides, which are usually eco-friendly, are more conducive to environmental sustainability and more beneficial to public wellness. Many species have anti-insect properties but are non-toxic to humans, including *Arisaema jacquemontii*, which has been demonstrated to have an anti-cancer potency.

Elimination of Breeding Grounds

Proper waste management and drainage of still water, eliminates the breeding ground of many pests.

Garbage provides food and shelter for many unwanted organisms, as well as an area where still water might collect and be used as a breeding ground by mosquitoes. Communities that have proper garbage collection and disposal, have far less of a problem with rats, cockroaches, mosquitoes, flies and other pests than those don't.

Open air sewers are ample breeding ground for various pests as well. By building and maintaining a proper sewer system, this problem is eliminated.

Poisoned Bait

Poisoned bait is a common method for controlling rat populations, however is not as effective when there are other food sources around, such as garbage. Poisoned meats have been used for centuries for killing off wolves, birds that were seen to threaten crops, and against other creatures.

Field Burning

Traditionally, after a sugar cane harvest, the fields are all burned, to kill off any insects, or eggs, that might be in the fields.

Hunting

Historically, in some European countries, when stray dogs and cats became too numerous, local populations gathered together to round up all animals that did not appear to have an owner and kill them. In some nations, teams of rat catchers work at chasing rats from the field, and killing them with dogs and simple hand tools. Some communities have in the past employed a bounty system, where a town clerk will pay a set fee for every rat head brought in as proof of a rat killing.

Traps

Traps have been used for killing off mice found in houses, for killing wolves, and for capturing raccoons and stray cats and dogs for disposal by town officials.

Poison spray

Spraying poisons by planes, hand held units, or trucks that carry the spraying equipment, is a common method of pest control. Throughout the United States of America, towns often drive a town owned truck around once or twice a week to each street, spraying for mosquitoes. Crop dusters commonly fly over farmland and spray poison to kill off pest that would threaten the crops. Many find spraying poison around their yard, homes, or businesses, far more desirable than allowing insects to thrive there.

Destruction of infected plants

Forest services sometimes destroy all the trees in an area where some are infected with insects, if seen as necessary to prevent the insect species from spreading. Farms infested with certain insects, have been burned entirely, to prevent the pest from spreading elsewhere.

Natural rodent control

Several wildlife rehabilitation organizations encourage natural form of rodent control through exclusion and predator support and preventing secondary poisoning altogether.

Eliminate Food Sources: Keep bulk food, seed, and dry pet food in metal cans with secure lids. Pick up fallen fruit. Take birdfeeders inside at night.

Remove potential rodent homes like yard debris, trash, construction waste, etc.

Exclude rodents from your home. Seal openings 1/2 inch or larger around the outside of your house with metal, concrete, or Copper Mesh Wool, which can be found online or at hardware stores.

Include natural rodent predators in your solution. A family of five owls can consume up to 3000 rodents in breeding season. Placing a nest box to encourage a family of owls to make your property home can be a great alternative to commercial pest control methods.

Use catch-and-release traps as a safe, sanitary, and humane solution. Catch-and-release traps will allow you to remove rodents from inside your home, but you must prevent their return by sealing entrance and exit holes and removing attractants (see above). Do not release the animal in an area that is unknown to the animal.

INTEGRATED PEST MANAGEMENT

Integrated Pest Management (IPM) brings together into a workable combination the best strategies of all control methods that apply to a given problem created by the activities of pests. IPM has been defined in various ways but a more scientific definition describes it as, "the practical manipulation of pest populations using sound ecological principles to keep pest populations below a level causing economic injury". The emphasis here is "practical" and "ecological". There are many ways of controlling insect pests but only a few are practical, and fewer are ecologically sound, such that an undesirable citation is created. Another term we frequently encounter is "intergrated pest control". It is offer used interchangeably with IPM, though in the strictest sense these terms are not identical.

Originally, integrated control simply meant modifying chemical control in such a way as to protect the beneficial insects and mites, or integrating chemical and biological control methods. Subsequently the concept was broadened to include all suitable methods that could be used in complementary ways to reduce pest populations and keep them at levels which did not cause economic damage. This essentially is IPM It includes a variety of options, any one of which may not significantly reduce the Pest population, but the sum total of which will give adequate reduction to prevent economic losses.

A modern definition of IPM may be-the use of all available tactics in the design of a program to manage, but not eradicate pest population so that economic damage and harmful environmental side effects are minimised. IPM is not a static, unyielding system. It is dynamic, ever-changing, as we develop a better understanding of all factors that affect the system. These factors include climate, alternate host plants, beneficial insects and man's activities. In a narrow sense, IPM means the management of the few important pests generally found on our crops, but consciously or not it must include all insect pests, not only the "key" ones but also the secondary pests, which seldom do any harm. If this were not so, we might suddenly find some of these minor insect pests or even non-pests elevated to the status of serious insect pests because of our failure to consider them in the total scheme.

IPM as a concept is not new, but one that is receiving new emphasis as man looks for better methods to grow and store food for an expanding population, and at the same time preserve his environment. The rationale for using IPM is threefold. First, it can cut production costs mainly by reducing energy inputs. Secondly, IPM can reduce environmental contamination through the judicious use or reduced use of pesticides. And finally, an IPM program allows for maximum utilisation of cultural practices and natural enemies and physical methods. IPM can be designed to take advantage of the ecological principles governing pest population abundance. This requires a thorough understanding of the role of all the factors responsible for a pest population reaching certain levels at a particular time of the year, or duration of storage.

Elements of IPM

There are four basic elements of IPM: natural control, sampling economic levels, and insect biology and ecology. The first element of IPM relates to the fullest utilisation of naturally occurring suppressive factors, including any practice by man which will make the total ecosystem less favourable for growth of the insect pest population. Obviously, this requires a thorough understanding of the ecosystem.

The naturally occurring suppressive factors may act directly or indirectly on pest populations. Indirectly, the ecosystem may be managed or altered in such ways as to make the environment more harmful to the pest and thus limit population growth. More directly, protection and the use of beneficial insects may help keep potentially damaging insect pest populations at subeconomic levels In storage, parameters that can be manipulated to control the buildup of a pest population are temperature, relative humidity, moisture content and composition of gases within the storage atmosphere. The second element is that of using sound economic threshold (ETL) levels as the basis for applying control measures, especially chemical measures. Establishing and using dynamic ETL's provide a basis for delaying the use of insecticides. This permits the maximum utilisation of other control methods, such as the use of beneficial insects.

The use of economic threshold levels implies adequate sampling of all harmful and beneficial insects in the agroecosystem and particularly in any one crop at a given time. The levels found through sampling must then be measured against the economic level established for the crop, the beneficial insects, and the probable population trend of the pest species. The sampler thus becomes a key person in an IPM system.

The fourth element, insect biology and ecology, is essential to the fullest utilisation of the other three elements. Little concerning natural control can be understood without detailed knowledge of the biology and ecology of all the species present. This knowledge is also essential in establishing the role of each species in the system and in deermining the amount of damage inflicted by each pest species. Adequate sampling is directly dependent a thorough familiarity of the species involved.

Knowledge of the biology of a certain problem pest will serve as a basis for planning the control strategies and provide operational guidelines for these strategies. In this context, it is important to know the relationship between the pest and the crop and the mortality factors, both biotic and abiotic which play a major role in the determination of pest population dynamics.

An understanding of the sequential dominance of pests in relation to growth stages could provide the immediate impetus for developing a simple integrated control program based on minimum pesticide application. By delineating the succession of major pests at different stages of plant growth, the frequency, timing and dosage of insecticide application could be synchronised, hence avoiding pesticide use on a time-wise basis, or the "calender" method. The control program could then be based on expected pest population at any given growth stage of storage duration.

There is a wide variation in the degree to which pests may be tolerated even for the same species in different areas, in different times of the year on different host plants, and in different stages of crop development. Thus, determination of the Economic Injury Level is critical in defining the ultimate aim of any pest management program, and in delineating the pest population level below which damage is tolerable and above which specific intervention is needed to prevent a pest explosion and to avert significant damage. Stern *et al.* defined Economic Injury Level as "the lowest pest population that will cause economic damage" while economic threshold level (ETL) of more accurately Control Action Threshlod (CAT) as "the density at which control measures should be applied to prevent an increasing pest population level from reaching the economic injury level". Although the damage or losses at the economic threshold can be tolerated of neglected, it is at this level that every effort should be made to reduce the pest population by various methods.

Determining the Economic Injury Level and Control Action Threshold is generally a complex matter based on detailed operations of pest ecology as it relates to bioclimatology, predation diseases, the effect of host plant resistance and the environmental consequences of applied control interventions. Some suggested that the following factors as essential for the determination of the Economic injury Level:

1. Amount of physical damage related to various pest densities.
2. Monetary value and production costs of the crop at various levels of physical damage.
3. Monetary loss associated with various levels of physical damage.
4. Amount of physical damage that can be prevented by the control measure.
5. Monetary value of the portion of the crop that can be saved by the control measure.

From this information, it is possible to determine the level of pest density at which control measures can be applied to save crop equal to, or exceeding the cost of control. The rise and fall of the Control Action Threshold is determined by the importance of the ecosystem, value of the crop, the pest status, and consumer standards.

Components of IPM

Five general types of single component control methods may be used in IPM programs in stored ecosystems. These are: chemical control, physical and mechanical methods, biological control, host plant resistance and regulatory control.

Chemical control

A variety of insecticides and acaricides have been and are continuously being developed for control of insect pests. However, these chemicals are but one tool and should be used in combination with other tactics in an IPM program. The total reliance on chemicals has led to a crisis situation. However, IPM does not advocate the complete withdrawal of pesticides. That would be impractical. IPM simply demands use of pesticide only when necessary and at rates compatible with other strategies.

Physical and mechanical methods

Physical and mechanical methods are direct or indirect measures that completely eliminate pests, or make the environment unsuitable for their entry, dispersal, survival and reproduction. Physical-mechanical control measures may include environmental manipulation, mechanical barriers, light taps, irradiation, thermal disinfestation, sanitation, etc.

Many times, mechanical and physical methods require considerable extra equipment, materials and labor, hence, they may only be economical in certain situations. For field pests, these methods are rather inefficient but in a storage ecosystem, many of the physical techniques are effective and have great potential for use in an IPM system.

Biological control

Biological control may be defined in a narrow sense as "the manipulation of predators or patho. yens to manage the density of an insect population". This definition does not include the naturally occurring control agents, but only parasitoids, predators and pathogens that are purposely manipulated by man. In a broader sense, it includes "the manipulation of other biological facets of the pest life system, such as its reproductive processes, its behavior, the quality of its food and so forth."

There are some constraints to the potential use and success of natural enemies. Predators, parasites and pathogens found amongst the grain will be regarded as contaminants by consumers and grain exporters. Thus, it makes it very difficult to maintain a pest population level that will enable the biological control agents to establish themselves. The use of pheromones is one of the potentially useful biological agents that could be utilised in IPM for monitoring and partially suppressing pest population not only in agricultural fields but in storage ecosystems.

Host-plant resistance

The manipulation of the genetic make up of the host so that it is resistant to pest attack is called host plant resistance. Over the years there have been numerous successes in breeding for resistance to a variety of pests and currently many crops are being selected for this purpose. This approach has not been attempted to any great extent in stored products protection systems Investigations in this field have been few. However, research has provided evidence of the utility of varietal resistance in grain storage. Unless research on varietal resistance to storage pests is integrated with breeding of plants that are resistant to field insect pests and deceases the potential of this tactic in storage IPM is limited.

Regulatory control

Fundamental regulatory control principles involve preventing the entry and establishment of foreign plants and animal pests in a country or-area, and eradicating, containing or suppressing pests already established in limited areas. Under the auspices of various quarantine acts, numerous control measures are implemented in an attempt to exclude potential pests, to prevent spread and to supplement eradication programs.

Ports of entry are the first line of defense against the introduction of new pests. Pests which break through the port of entry are eradicated or contained within limited areas.

Quarantine action is used only against insects of economic importance, although it is sometimes necessary to contain insects which are of no economic importance in another country until their behavior in a new environment can be studied. Trogoderma granarium is a most serious pest of stored commodities and every effort is extended to prevent its spread in international trade. In many countries, imported consignments found to contain J. granarium are segregated and immediately fumigated with methyl bromide. Lately, Prostephanus truncates originating from Central America has become a pest of international quarantine importance.

Component integration

Each of the many methods in insect control has its place in IPM. There are many situations where two or more can be used in an integrated program. Not all methods, however, are suitable for use in every situation.

In a storage ecosystem, hygiene and good warehouse management are essential. It provides the framework for other supplementary infestation control methods. An IPM system would therefore supplement sanitation and good warehouse keeping with one or more combination of the following practices:

1. Improved harvesting and threshing techniques
2. Judicious use of residual insecticides
3. Use of fumigants (MeBr; PH_3)
4. Use of ambient aeration, and refrigerated aeration
5. Atmospheric gas modification (hermetic; CO_2; N_2)
6. Thermal disinfestation
7. Irradiation techniques
8. Insect resistant packaging
9. Insect growth regulators: (IGRs: methoprene, hydroprene)
10. Biological control (parasites, predators and entomopathogens, pheromones)
11. Use of resistant varieties if possible
12. Storage management (FIFO)
13. Adequate grain cleaning prior to storage storage.
14. Storage design (for pest exclusion, principally for rodont and bird pests)
15. Adequate grain cleaning prior to storoge

16. Monitoring, evaluation and inspection of stored commodities, storage structures and their immediate surroundings.

Insect Pest Management for Stored-Products

There is a number of differences for IPM for stored-products compared to field agriculture. There is far greater variety of tactics that could be employed for storage pest management. In both situations, a systems approach is usd in order to facilitate monitoring and implementation.

Comparison of IPM between field agriculture and stored-product pests

Field production

1. Ecological condition—more complex and dynamic.
2. Sanitation process—difficult to implement.
3. Economic threshold—appropriately applicable.
4. Physical control—often impractical and expensive.
5. Mechanical control—often impractical and expensive.
6. Regulatory control—more complex and require intensive logistical support.
7. Chemical control—considerable impact on the ecosystem.
8. Biological control—field
9. Host—plant resistance—whole plant modification relatively easier.

Stored—product setting

1. Relatively simple and manageable.
2. Easier and alone could provide complete control.
3. Not generally applicable, food industry often require zero infestation.
4. Most physical factors could be manipulated under storage condition.
5. More feasible by as built—in feature of storage.
6. Detection and implementation generally selfcontained.
7. More contained and limited in spatial proliferation.
8. Problem of contamination.
9. Genetic manipulation to render resistance to seed only more complicated.

In both situations three additional elements must be incorporated to the development of IPM: 1) appropriate people, 2) a systems approach, and 3) adequate evaluation. Aside from the individual expertise of the team members it is important that the people involved must develop a good personal and working relationship in the pursuance of the common objectives.

ECONOMIC PRINCIPLES IN PEST MANAGEMENT

Entomologists and biologists defing pests as those organisms reaching a certain biomass or population level enough to cause physiological damage to crop and eventually impair either the quantity and quality of the produce. The concern used to be concentrated solely on devising control strategies with the aim of increasing the pest-kill efficiency and/or minimising the control's unwanted side effects on the environment. In doing so a damage function and a kill-function are determined simultaneously. Perhaps, the ultimate goal of these efforts is to realise the yield potential of crops by eliminating pest hazards which is possible only where there is maximum protection.

In years past, maximum protection was closely associated with prophylactic dosages of pesticides. Economists view pest problems and solutions quite differently. For one, pests are those organisms that destroy what man has produced, compete for resources needed by man in the production of food and fiber, or contaminate the environment in such a way that man finds it unhealthy or unattractive.

An economist would not be very particular about the exact levels of pests, damages, and mortalities; but would pay more attention to the value of pest damage and the costs of alternative control strategies with or without special consideration to time. Usually, he would not desire for the yield potential of crops because he knows that it does not always pay to spend for maximum protection. In fact, the urge to control pests does occur to him until he finds that the cost of doing so is more than offset by the benefits.

Biologists, to a great extent, do not worry about costs in as much as they find delight in quantifying apparent yield increases due to control action. Almost for two decades now, attempts have been made to reconcile the disciplinary concerns of both the natural and social scientists to benefit the farmers and the public at large. To put their interests together, the US National Academy of Sciences states that: entomologists develop and perfect various methods of insect control and determine their physiological impacts on crops; economists depend on entomologists for information regarding various inputs combined to produce a particular outcome. To some extent, pest management was conceived to pool together the concerns of people coming from various fields in a most coordinated fashion.

Pest management is defined as an ever-changing process of attacking pest problems by applying whatever economically feasible control of combination of controls that produces the best combination of immediate and long-term results in terms of both reduced pest damage and the absenec of unwanted side effects.

Input/cost. The factors of production such as buildings, machines, materials, labor and management skills are called inputs and if valued at their respective prices are called farm costs or expenses. Let x = quantity of input and Px = Price of input. Then Cost = (Px)(X)

Output/beturn. Through some physical or biological processes, the above inputs are transformed into usable products or output. Similarly, it can be valued by its price to get the return, revenue or benefit. The benefits from pest control is indicated by the increase in yield or added returns.

Let X = yield (increase) and Py = price of corp. Then Return or Revenue = (Py)(Y)

Pest management. Pest management is becoming more essential in our modernising agriculture as problems about pests and damages increase with time. As an input it has 5 components: pest control materials, labor and management or knowledge about pests, expected damage, and alternative control methods.

Pesticides. Pesticides used to be the single most important protectors of crops from insects, weeds and pathogens but because of the ever-increasing price and unwanted side effects, nonchemical methods are being tapped. Nevertheless, continued dependence on pesticides in the near future is a reality that pest managers have to reconcile with. Generally, pesticides form a regular feature of the farm cash expenses.

Non-chemical control. This method is likewise called labor control because it relies heavily on a farmer's manual efforts. Non-chemical methods do not necessarily mean that the materials needed for its utilisation are cost-free.

Efficiency. We define the most efficient pest control strategy as one that requires the least combination of labor cash input for a given level of yield increase, their factors held constant. The definition combines the "efficiency" and effectiveness concepts presented by Semple. According to him the efficiency of an insecticide is the minimum dosage required to reduce the population of target pests by a fixed proportion (80%) regardless of pest pressure. Effectiveness, on the other hand, is indicated by the number of surviving insects of the extent of yield damage after the treatment. It is determined not by the "efficiency" of insecticides but by the strength of pest perssure as insecticides by the over-all farm operations. One needs to know the compatibility of his control measure to his pest problems and an effective manager does.

Optimum thresholds. At least, there are two sets of pest population thresholds found in pest management literature, one being a shadow of the other. Studies regarding plant physiology in relation to phytophagous organisms that reveal there is a tolerable level of plant destruction due to pests wherein the crop is still able to recover and remain vigorous. Apparently, economic injury level (EIL), sometimes termed as the biological threshold level (BTL) is the critical pest density level going beyond what is tolerable to crops. With the EIL, biologists justify control actions. For the economists, such could never be unless the cost of control is zero.

Economic threshold level (ETL) is the level of pest density at which suppression action is initiated, the control cost at least equals the value of expected damage. Some IPM specialists prefer to use the term control action threshold (CAT) for EIL since it connotes immediately that it is that level control action is necessary. The mathematical expression for ETL below tells the basic relationship between the critical pest density level and economic variables. If we assume d and K as pre-determined variables then only the cost and crop prices affect the ETL. Note the prohibitive character of cost control action. If prices of pesticides would be very, very high, ETL would also be and control action may not be economically warranted. However, if the crop is of high value or if the market requires high quality products, ETL will be small and the propensity for control action will be high.

$$\text{ETL} = c/pdk$$

Where

c = cost of control action

r = price of crop

d = damage coefficient

k = kill efficiency

Partial Budgetting

A private farmer, faced with a limited cash budget and credit would think twice before he puts in money into an otherwise biologically attractive innovation. He might not be doing detailed accounting of the proposed change but deep in his mind, he calculates and balances the relative benefits and costs of every new undertaking. If the perceived economic net gain is positive and acceptable, then he evaluates the resource requirements and explores ways finance it.

Partial budgetting is perhaps the most popular decision tool employed in the conduct of applied research by the entomologists in an attempt to add economic substance to experimental results. Partial budgets are designed to show, net profit or loss for the farm as a whole, but the net increase in farm income resulting from a minor change in the on-going operations. It contains 4 major items bearing exact information about the changes in monetary terms, namely: added returns, reduced costs, added costs, and reduced returns. Added returns is the value of the increase in yield and/or the value of quality improvements.

Reduced costs is the value of inputs no longer used under the proposed change. If previously the farm is using 6 man-days of labor but the new technology requies only 4 man-days, then the wage that should have been paid to 2 man-days represent a reduction in cost.

Added costs is the value of additional inputs as a result of the proposed change. This is the exact opposite of reduced costs. The classical example is the cost of additional fertilisation, labor and machine service associated with modern farming. Reduced returns is the value of forgone output. It the proposed change diminishes yield, then it reduces returns. In some instances, the increase in production is shared with the landlord and harvester/thresher so that the value of harvest that went to them is accounted as reduced returns.

The procedure for partial budgetting is straightforward. Analysts have just to be keen the details in the change of process. Given data on market prices, he can put values to each change item and determine if such diminishes or adds to returns or costs. After which, he can compute for the net change in income by subtraction the sum of Added costs and Reduced returns from the sum of added returns and reduced costs:

Net gain/loss in income = (Added returns + Reduced Costs)—(Added costs + Reduced returns)

Biological studies on pest control strongly assume that farmers do not control pests; this is the implication of comparing- treated as against the untreated plots. Such standard experimental procedures directly ease out computations involved in the economic analysis but, at the expense of intuitive value of practical content. It does not provide the analyst a grasp of the general and actual problem situation because at this time very few farmers are indifferent to pests. Pesticide consumption growths observed for rice, vegetables and bananas indicate that pest control is very active in our agricultural/farming system. Quite recently, field studies on the control of rice pests use farmers' practice as a separate benchmark for comparison.

It is quite necessary that analysts look for flexibilities in the data set and mold the framework according to the dictates of the local problem situation. In that case, he could come up with a matrix of problem situations and alternatives. For each situation, the dominant alternative may be decided based on at least 2 criteria: (1) magnitude of net change in income (2) resource requirements and accessibility.

In reality, farms are multi-enterprise in nature. There is also substantial interdependency in decision across crops and over time. Uncertainties or lack of information about pests, damage, controls and prices also exist. In other words, the type of control a farmer applies to his main crop influences his control decision for the second crop. Similarly, his control practices in the previous season greatly affect his decision in both present and future cropping.

Imperfect knowledge and limited skills concerning pest control decisions and the presence of natural disturbances may render actual decisions "inferior" to that predicted by the partial budget model. And as one tries to develop realistic models, the analytical tools tend to be conceptually complicated and computationally irksome. This leads us to a brief discussion of the various models developed recently.

Whole farm budgets. This is merely an extension of the partial budgets. It recognises interdependent decisions which means that an adoption of technology in one enterprise affects the rest. Thus, the entire farming system is changed. Oftentimes, it will be hard to distinguish the cause from the effect. Taking the entire system as the unit for analysis makes it simpler.

Production function analysis. It is basically statistics in form and is widely used in production economic research. A production function tells the average relationship between the farm product and its factors of production. It bears powerful estimates of the precise contribution of each input to the total product. Headley formalised the economic study of pest control and applied this framework in the analysis of pesticide expenditure in US agriculture. He concluded that pesticides are indeed productive inputs placing a $4 contribution to the value of agricultural production per dollar spent on it. He thus predicted an expanding market for agricultural pesticides in the 70's and challenged the technical faculty to develop a less hazardous but comparatively productive substitute.

Benefit-cost analysis. BCA is probably the most comprehensive analytical tool unfortunately, it requires voluminous data and it is usually not workable. BCA treats pest control as an investment or a folw of costs and returns. Though time is the core of B-C analysis wherein discounting methdods are used to calibrate the flow of money in order to come up with timeless measures of efficiency the internal rate return (IRR) and the benefit-cost ratio (BCR). The ideal situation is for the IRR to be greater than the market interest rate and for the BCR to be greater than 1. BCA's can be done both at the farm and the community levels. For the latter, we call it extended BCA.

Risk analysis. Finally, the latest add-on to our list views the resurgence of pests as an involuntary risk taken by farmers, pest control actions as means to minimise the risks or variability of yields and profits, and control costs as the premium farmers pay to avoid it. This analysis exerts less pressure on biological data requirements, specifically, pest infestation levels. This type of analysis has the potential of influencing pest control policies particularly crop insurance and pest information services.

REFERENCES

Bennett, Gary W., et.al. 2005. *Truman's Scientific Guide to Pest Management Operations.* Purdue University, Questex.

Jahn, GC, JA Litsinger, Y Chen and A Barrion. 2007. "Integrated Pest Management of Rice: Ecological Concepts". In *Ecologically Based Integrated Pest Management.* CAB International Pp. 315-366.

Kogan, M 1998. *Integrated Pest Management:Historical Perspectives and Contemporary Developments,* Annual Review of Entomology Vol. 43: 243-270.

13

Harvesting of Horticultural Products

Effective management during the postharvest period, rather than the level of sophistication of any given technology, is the key in reaching the desired objectives. While large scale operations may benefit from investing in costly handling machinery and high-tech postharvest treatments, often these options are not practical for small-scale handlers. Instead, simple, low cost technologies often can be more appropriate for small volume, limited resource commercial operations, farmers involved in direct marketing, as well as for suppliers to exporters in developing countries.

Many recent innovations in postharvest technology in developed countries have been in response to the desire to avoid the use of costly labor and the desire for cosmetically "perfect" produce. These methods may not be sustainable over the long term, due to socioeconomic, cultural and/or environmental concerns. For example, the use of postharvest pesticides may reduce the incidence of surface defects but can be costly both in terms of money and environmental consequences. In addition, the growing demand for organically produced fruits and vegetables offers new opportunities for small-scale producers and marketers.

Local conditions for small-scale handlers may include labor surpluses, lack of credit for investments in postharvest technology, unreliable electric power supply, lack of transport options, storage facilities and/or packaging materials, as well as a host of other constraints. Fortunately, there is a wide range of simple postharvest technologies from which to choose, and many practices have the potential of meeting the special needs of small-scale food handlers and marketers. Many of the practices included in the manual have successfully been used to reduce losses and maintain produce quality of horticultural crops in various parts of the world for many years.

There are many interacting steps involved in any postharvest system. Produce is often handled by many different people, transported and stored repeatedly between harvest and consumption. While particular practices and the sequence of operations will vary for each crop, there is a general series of steps in postharvest handling systems that will be followed for the purposes of the manual.

Despite decades of educational efforts, the most common causes of postharvest losses in developing countries continue to be rough handling and inadequate cooling and temperature maintenance. The lack of sorting to eliminate defects before storage and the use of inadequate packaging materials further add to the problem. In general, minimizing rough handling, sorting to remove damaged and diseased produce and effective temperature management will help considerably toward maintaining a quality product and reducing storage losses. Storage life will be enhanced if the temperature during the postharvest period is kept as close to the optimum as feasible for a given commodity.

HARVESTING PRACTICES

Small-scale producers have the option to harvest earlier, when vegetables are more delicate and valuable; harvest later, when fruits are at a riper, more flavorful stage; or harvest more often (taking advantage of multiple harvests to gather produce at its optimum stage of maturity). All these options can lead to higher profits due to the higher value of the produce you have to offer for sale.

One of the most common mistakes growers make is to harvest fruit crops too early, when they are under-ripe and have not yet developed their full flavor. Some vegetables, if allowed to grow large, will be too fibrous or full of seeds for good eating quality. With many horticultural crops, if you harvest all at once you are sure to have many items that are either under-mature or over-mature. Using a maturity index as a standard will greatly reduce pre-sorting losses. For some crops this involves using a refractometer to measure sugars or a penetrometer to measure firmness.

Mechanical damage during harvest can become a serious problem, as injuries predispose produce to decay, increased water loss and increased respiratory and ethylene production rates leading to quick deterioration. In general, harvesting by machine will cause more damage than harvesting by hand, although some root crops can be severely damaged by careless hand digging. The containers used by pickers in the field should be clean, have smooth inside surfaces and be free of rough edges. Stackable plastic crates, while initially expensive, are durable, reusable and easily cleaned. If baskets must be used, they should be woven "inside out" with the stubs of the beginning and end of each cane on the outside of the basket.

Manual harvesters should be well trained in the proper way to harvest the crop to minimize damage and waste, and should be able to recognize the proper maturity stage for the produce they are handling. Pickers should harvest with care, by snapping, cutting or pulling the fruit or vegetable from the plant in the least damaging manner. The tips of knives should be rounded to minimize inadvertent gouges and excess damage to perennial plants. Knives and clippers should always be well sharpened. Pickers should be trained to empty their picking bags and/or baskets with care, never dumping or throwing produce into field containers. If harvesters pick directly into large bulk bins, produce can be protected from bruising by the use of a de-accelerating chute fashioned from canvas. Vented, stackable field containers should be kept clean and smooth.

Exposure to the sun should be avoided as much as possible during and after harvest, as produce left out in the sun will gain heat and may become sun-burned. Produce exposed to sunlight can soon become 4 to 6 °C (7 to 11 °F) warmer than air temperature. Field bins should be placed in the shade or loosely covered (for example with light colored canvas, leafy plant materials, straw or an inverted empty container) if delays are expected in removing them from the field. Night or early morning harvest is sometimes an option for harvesting produce when internal temperatures are relatively low, reducing the energy needed for subsequent cooling. Latex flow is often lower later in the morning than it is at dawn for crops such as mango and papaya), so harvesting as late in the morning as possible can reduce later efforts required to clean the produce before packing. Also, citrus fruits should not be harvested early in the morning when turgid because of their greater susceptibility to the release of essential oils from the flavedo oil glands that cause oil spotting (green spots on yellow and orange citrus fruits after degreening).

Directly following harvest, when produce is prepared for marketing, cooling is essential. Cooling (also known as "pre-cooling") is the removal of field heat directly after harvest, before any further handling. Any delays in cooling will shorten postharvest life and reduce quality. Even produce undergoing repeated cooling and warming deteriorates at a slower rate than produce that has not been cooled.

Rough handling during preparation for market will increase bruising and mechanical damage and limit the benefits of cooling. Roads between the field and the packinghouse should be graded and free from large ruts, bumps and holes. Field boxes must be well-secured during transport and, if stacked, not overfilled. Transport speeds must be suited to the quality and conditions of the roads, and truck and/or trailer suspensions kept in good repair. Reduced tire air pressure on transport vehicles will reduce the amount of motion transmitted to the produce.

Any practice that reduces the number of times the produce is handled will help reduce losses. Field packing (selection, sorting, trimming and packaging of produce at the time of harvest) can greatly reduce the number of handling steps the produce must undergo before marketing. Small carts or small mobile field packing stations can be designed to be moved along with the packers and to provide shade for packing operations.

MATURITY STANDARDS

Maturity standards have been determined for many fruit, vegetable and floral crops. Harvesting crops at the proper maturity allows handlers to begin their work with the best possible quality produce. Produce harvested too early may lack flavor and may not ripen properly, while produce harvested too late may be fibrous or overripe. Pickers can be trained in methods of identifying produce that is ready for harvest. The following table provides some examples of maturity indices.

Table 1. Maturity indices of fruit, vegetable and floral crops

Index	*Examples*
Elapsed days from full bloom to harvest	Apples, pears
Mean heat units during development	Peas, apples, sweet corn
Development of abscission layer	Some melons, apples, feijoas
Surface morphology and structure	Cuticle formation on grapes, tomatoes Netting of some melons; Gloss of some fruits
Size	All fruits and many vegetables
Specific gravity	Cherries, watermelons, potatoes
Shape	Angularity of banana fingers; Full cheeks of mangos; Compactness of broccoli and cauliflower
Solidity	Lettuce, cabbage, brussels sprouts
Textual properties	
Firmness	Apples, pears, stone fruits
Tenderness	Peas
Color, external	All fruits and most vegetables
Internal color and structure	Formation of jelly-like material in tomato fruits Flesh color of some fruits
Compositional factors	
Starch content	Apples, pears
Sugar content	Apples, pears, stone fruits, grapes
Acid content, sugar/acid ratio	Pomegranates, citrus, papaya, melons, kiwifruit
Juice content	Citrus fruits
Oil content	Avocados
Astringency (tannin content)	Persimmons, dates
Internal ethylene concentration	Apples, pears

Vegetables are harvested over a wide range of maturities, depending upon the part of the plant used as food. The following table provides some examples of maturity indices of vegetable crops.

Crop	*Index*
Root, bulb and tuber crops	
Radish and carrot	Large enough and crispy (over-mature if pithy)
Potato, onion, and garlic	Tops beginning to dry out and topple down
Yam bean and ginger	Large enough (over-mature if tough and fibrous)
Green onion	Leaves at their broadest and longest
Fruit vegetables	
Cowpea, yard-long bean, snap bean, batao, sweet pea, and winged bean	Well-filled pods that snap readily
Lima bean and pigeon pea	Well-filled pods that are beginning to lose their greenness
Okra	Desirable size reached and the tips of which can be snapped readily
Upo, snake gourd, and dishrag gourd	Desirable size reached and thumbnail can still penetrate flesh readily (over-mature if thumbnail cannot penetrate flesh readily)
Eggplant, bitter gourd, chayote or, cucumber	Desirable size reached but still tender (over-mature if color dulls or changes and seeds are tough)
Sweet corn	Exudes milky sap from kernel if cut
Tomato	Seeds slipping when fruit is cut, or green color turning pink
Sweet pepper	Deep green color turning dull or red
Muskmelon	Easily separated from vine with a slight twist leaving clean cavity
Honeydew melon	Change in fruit color from a slight greenish white to cream; aroma noticeable
Watermelon	Color of lower part turning creamy yellow, dull hollow sound when thumped
Flower vegetables	
Cauliflower	Curd compact (over-mature if flower cluster elongates and become loose)
Broccoli	Bud cluster compact (over-mature if loose)
Leafy vegetables	
Lettuce	Big enough before flowering
Cabbage	Head compact (over-mature if head cracks)
Celery	Big enough before it becomes pithy

Using a refractometer

Sugars are the major soluble solids in fruit juices and therefore soluble solids can be used as an estimate of sweetness. A hand-held refractometer can be used outdoors to measure % SSC (equivalent degrees Brix for sugar solutions) in a small sample of fruit juice. Temperature will affect the reading (increasing about 0.5% SSC for every 5 ºC or 10 ºF), so you should adjust the measurement for the ambient temperature.

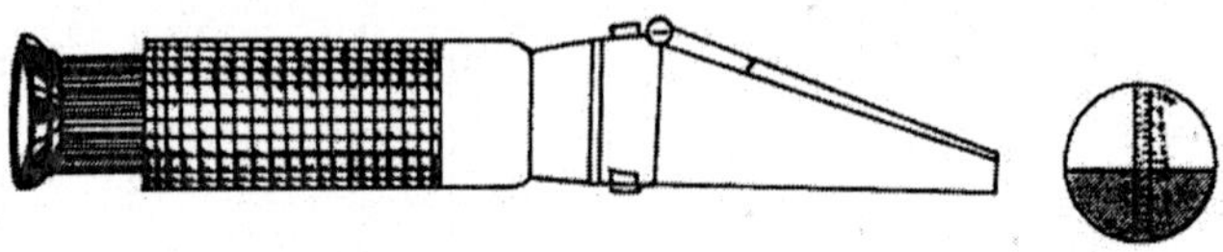

Figure 1. A hand-held refractometer

A garlic press works well to squeeze the juice from fruit samples. For small fruits, use the whole fruit. For large fruits, take a wedge for the stem end to the blossom end and to the center of the fruit. Remove any pulp by filtering the juice through a small piece of cheesecloth. You must clean and standardize the refractometer between each reading with distilled water (should read 0% SSC at 20 ºC or 68 ºF).

Here are some examples of proposed minimum % SSC for selected commodities. If your reading indicates a higher % SSC, then your produce is better than the minimum standard. Strawberries which are of excellent flavor, for instance, would measure 8% SSC or above.

Table 2. Minimum %SSC

Apricot	10%
Blueberry	10
Cherry	14-16
Grape	14-17.5
Kiwifruit	6.5
Mango	10-12
Muskmelon	10
Nectarine	10
Papaya	11.5
Peach	10
Pear	13
Pineapple	12
Plum	12
Pomegranate	17
Strawberry	7
Watermelon	10

Using a Firmness Tester

The degree of softness or crispiness can be estimated by squeezing produce, or by taking a bite. Objective measurements can be made with inexpensive penetrometers. The most common way to measure firmness is resistance to compression or pounds-force (lbf). The Effe-gi fruit penetrometer is a hand-held probe with a gauge for pounds-force.

To measure firmness, use fruit that are uniform in temperature, since warm fruit are usually softer than cold fruit. Use fruits that are uniform in size, since large fruit are usually softer than smaller fruit. Make two puncture tests per fruit on larger fruits, once on opposite cheeks, midway between stem and blossom ends. Remove a disc of skin (larger than the tip to be used) and choose the appropriate plunger tip. Hold the fruit against a stationary, hard surface, and force the tip into to fruit at a slow, uniform speed (take 2 seconds) to the scribed line on the tip. Take the reading to the nearest 0.5 lb-force.

Table 3. Appropriate Effi-gi plunger tip sizes to use when measuring firmness in selected fruits

1.5mm	(1/16 inch)	Olive
3 mm	(1/8 inch)	Cherry, grape, strawberry
8 mm	(5/16 inch)	Apricot, avocado, kiwifruit, pear, mango, nectarine, papaya, peach
11 mm	(7/16 inch)	Apple

Calibrate firmness testers by holding the tester vertically and placing the tip on the pan of a scale. Press down until the scale registers a given weight, then read the firmness tester. Repeat 3 to 5 times, if you find the instrument reads the same as the scale, it is ready to use. You can adjust the penetrometers by inserting washers in the appropriate locations (follow the instructions that come with the instrument).

HARVESTING PRACTICES

Harvesting practices should cause as little mechanical damage to produce as possible. Gentle digging, picking and handling will help reduce crop losses.

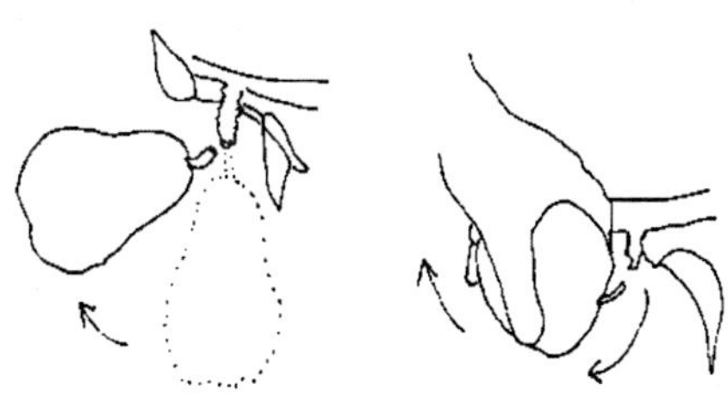

Figure 2. Pick carefully to avoid damage

For some crops, a natural break point forms at the junction of the stem and the stalk when produce is mature. Harvesters should grasp the product firmly but gently and pull upward as illustrated below. Wearing cotton gloves, trimming fingernails, and removing jewelry such as rings and bracelets can help reduce mechanical damage during harvest.

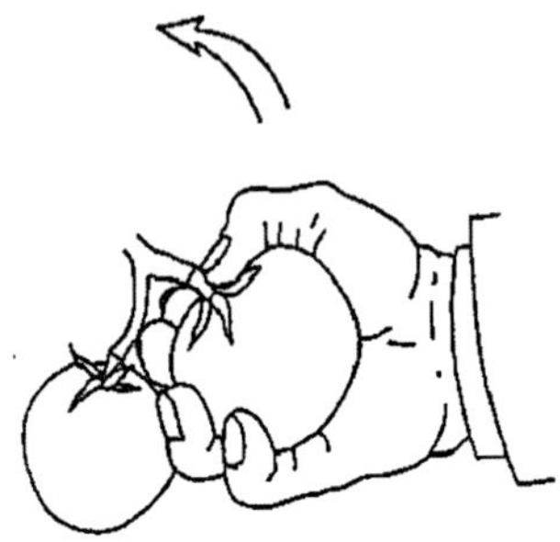

Figure 3. Gentle collection of tomatoes

If a small amount of leafy vegetables are being harvested for home use or for sale at a nearby roadside or farmers' market, a small tub of cold water can be useful for cooling the produce. The tub can be brought directly to the field and used by the picker as a field container. Clean water should be used with each lot of produce. Chilling leafy vegetables by using cold water at harvest will help maintain quality and prevent wilting.

Figure 4. Haresting of leafy vegetables

HARVESTING CONTAINERS

Picking baskets, bags and buckets come in many sues and shapes. These harvesting containers can be made by sewing bags with openings on both ends, fitting fabric over the open bottom of ready-made baskets, fitting bags with adjustable harnesses, or by simply adding some carrying straps to a small basket.

Plastic crates are relatively expensive but are durable, reusable and easy to clean. When empty, they can be nested to save space in storage or transport.

HARVESTING TOOLS

Some fruits need to be clipped or cut from the parent plant. Clippers or knives should be kept well sharpened. Penduncles, woody stems or spurs should be trimmed as close as possible to prevent fruit from damaging neighboring fruits during transport.

Pruning shears are often used for harvesting fruits, some vegetables, and cut flowers. A variety of styles are available as hand held or pole models, including shears that cut and hold onto the stem of the cut product. This feature allows the picker to harvest without a catching bag and without dropping fruits.

Figure 5. Straight bladed hand shears for fruits and flowers:

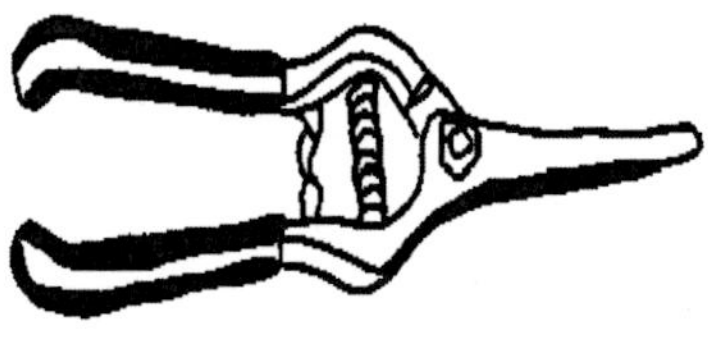

Figure 6. Thin curved blade for grapes and fruits

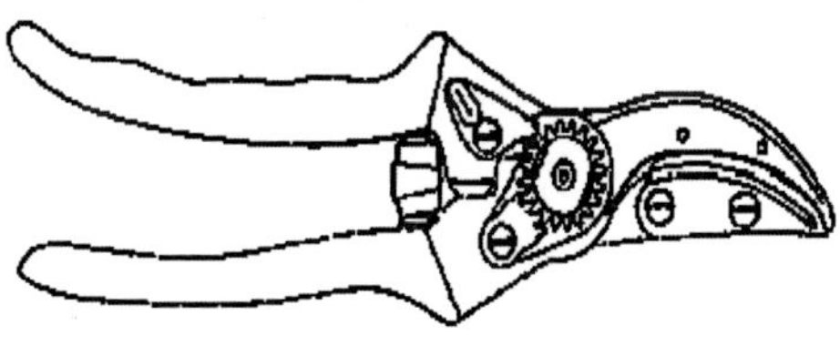

Figure 7. Cut and hold hand shears

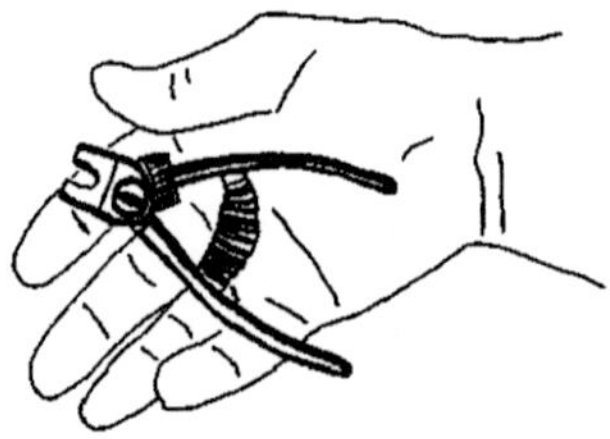

Figure 8. Clipper for citrus fruits

Figure 9. Pole mounted cut and hold picking shears

Using a cutting tool attached to a long pole can aid picking of crops such as mangoes and avocados when the fruit is difficult to reach. Cutting edges should be kept sharpened and the catching bag should be relatively small The angle of the cutting edge and the shape of the catching bag can affect the quality of the fruit harvested, so it is important to check performance carefully before using any new tools.

Figure 10. Using a picking pole

Picking poles and catching sacks can be made by hand or purchased from horticultural supply companies. The collection bags illustrated below were hand woven from strong cord or sewn from canvas. The hoop used as the basket rim and cutting edges can be fashioned from sheet metal, steel tubing or recycled scrap metal.

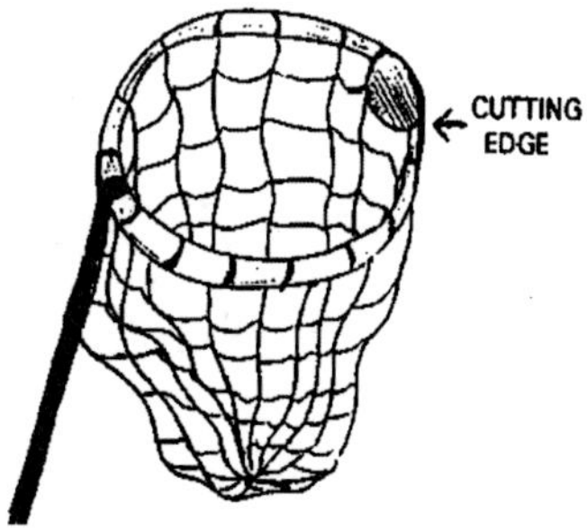

Figure 11. Hand woven collection bag

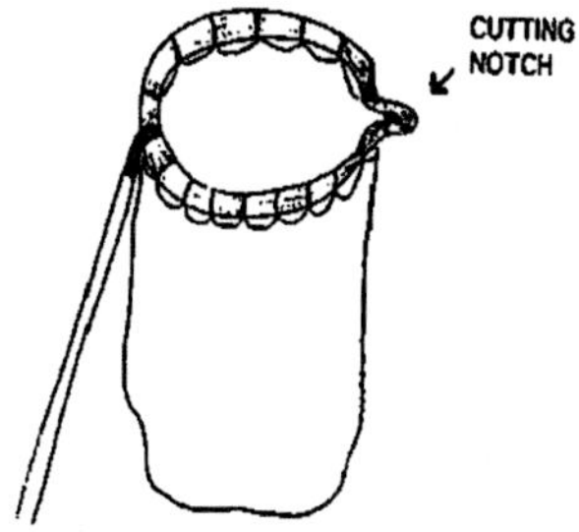

Figure 12. Canvas collection sack

Fruit trees are sometimes quite tall and letting fruit fall to the ground when it is cut from the tree will cause severe bruising. If two pickers work together, one can clip or cut the fruit from the tree, and the other can use a sack to break its fall. The catcher supports the bag with his hands and one foot, catches the falling fruit, then lowers the far end of the bag to allow the fruit to roll safely to the ground. Unlike most nut crops, pistachio nuts should not be knocked to the ground during harvest because of their open shells and relatively high moisture content. Plastic sheeting or canvas tarpaulins are spread below the tree being harvested, and trees are mechanically shaken or hand knocked (the branches hit with mauls) until the nuts drop.

FIELD PACKING

When crops are field packed the picker harvests and then immediately packs the produce after minimal handling Strawberries are generally field packed, since even a small amount of handling will damage these soft fruits. When lettuce is field packed, several wrapper leaves are left on the head to help cushion the produce during transport.

Figure 13. Cart for field packing

A small cart can help reduce the amount of bending and lifting the picker has to do during harvest. A simple aid for field packers is a movable cart with a rack for boxes and a roof to provide shade. This cart is designed to be pushed along the outer edge of the small field where harvest is taking place. This cart for field packing is designed to be pulled by a small tractor into the field when the crop is harvested. This type of cart can be used for field packing many types of crops. The roof folds down for easy transport, and opens up to provide a wide area of shade for the packers and the commodity. The cart design can be modified as needed to suit various products and different operations. A self-propelled field pack system allows field workers to cut, trim, tie/wrap and pack *in the field,* thus eliminating the expense of operating a packing shed. When crops are harvested at some distance from the packinghouse, the produce must be transported before packing.

CURING ROOT, TUBER AND BULB CROPS

Curing root and tuber crops such as sweetpotatoes, potatoes, cassava and yarns is an important practice if these crops are to be stored for any length of time. Curing is accomplished by holding the produce at high temperature and high relative humidity for several days while harvesting wounds heal and a new, protective layer of cells form. While curing can be initially costly, the long extension of storage life makes the practice economically worthwhile. The best conditions for curing vary among crops as shown in the following table:

Table 5. The best conditions for curing crops

Commodity	*Temperature*		*Relative Humidity*	*Days*
	°C	°F	(%)	
Potato	15-20	59-68	90-95	5-10
Sweetpotato	30-32	86-90	85-90	4-7
Yarns	32-40	90-104	90-100	1-4
Cassava	30-40	86-104	90-95	2-5

Curing, when used for onions, garlic and flowering bulbs refers to the practice directly following harvest, of allowing the external layers of skin and neck tissue to dry out prior to handling and storage. If local weather conditions permit, these crops can be undercut in the field, windrowed and left there to dry for five to ten days. The dried tops of the plants can be arranged to cover and shade the bulbs during the curing process, protecting the produce from excess heat and sunburn. If forced heated air is used for curing onions and other bulbs, one day or less at 35 to 45 °C (95 to 113 °F) and 60 to 75% relative humidity is recommended. The dried layers of 'skin" then protect the produce from further water loss during storage.

Field curing

Yams and other tropical root and tuber crops can be cured outdoors if piled in a partially shaded area. Cut grasses or straw can be used as insulating materials and the pile should be covered with canvas, burlap or woven grass mats. Curing requires high temperature and high relative humidity, and this covering will trap self-generated heat and moisture. The stack should be left for about four days

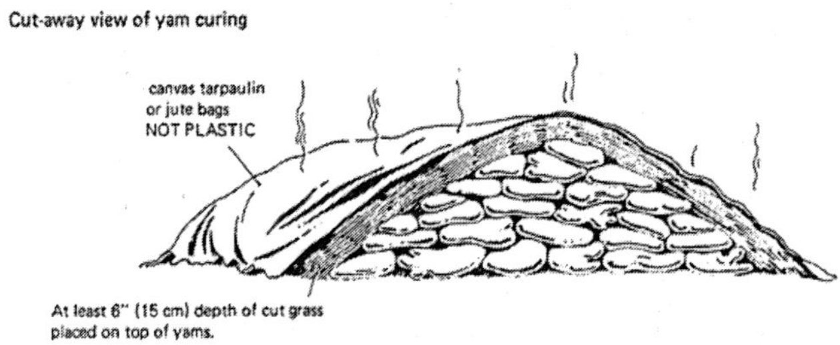

Figure 14. Cut-away view of yam curing

Onions and garlic can be cured in the field in regions where harvest coincides with the dry season. The crops can be cured either in windrows or after packing into large fiber or net sacks. The produce can be left in the field for five days, then checked daily until the outer skin and neck tissues are properly dried. Curing may take up to ten days, depending on weather conditions.

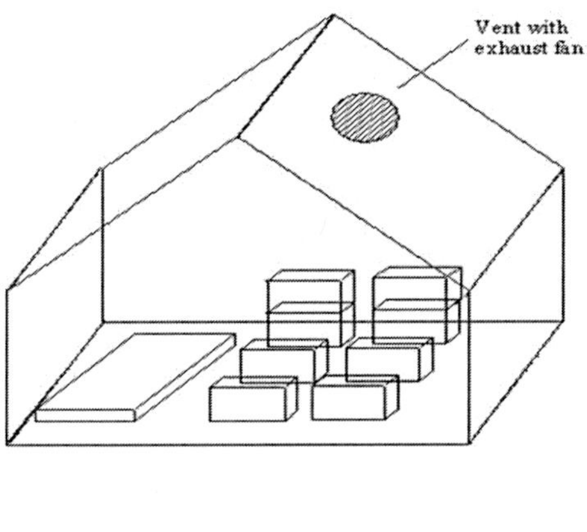

Figure 15. Curing assisted by shade and ventilation

Curing can be assisted by the use of ventilated sheds in regions where solar radiation and/or relative humidity is high or natural air movement is low. Produce in sacks can be stacked in the shade on canvas tarpaulins, or placed in an open sided shed under one or more ceiling fans. An exhaust vent in the roof can assist with air circulation.

Curing with heated air

The most uniform distribution of heat is obtained when heat is introduced near the floor level of a curing structure. Heaters can be placed on the floor near the bins of produce, or heat can be ducted in from outside the curing room. A high relative humidity can be obtained by wetting the floor or by using an evaporative cooler in the room without introducing outside air.

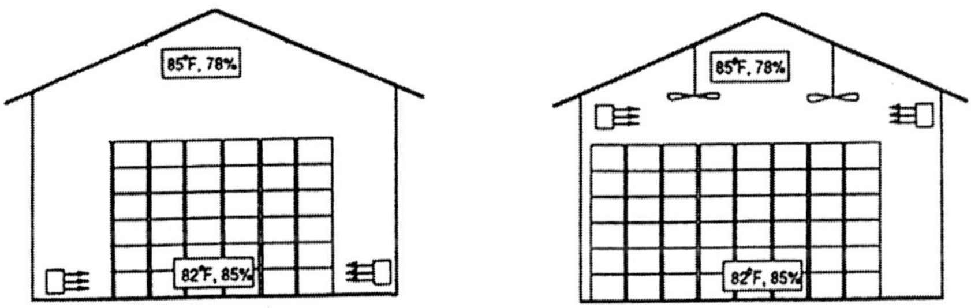

Figure 16. Curing with heated air

If heaters are located near the ceiling, then ceiling fans can be used to help redistribute the heat down into the room of produce. Bulk bins must be stacked to allow a gap of 10 to 15 cm (4 to 6 inches) between rows for adequate air circulation.

Bulk systems for curing onions

Curing using a bulk system requires a fan, a heating unit and a slatted floor. The illustrations below shows how air can be brought in, heated and distributed through a load of bulk onions in the curing room An exhaust opening near the ceiling re-circulates heated air. When using heated air it is easy to over-dry the bulbs, leading to loss of external scales and exposure of the fleshy scales underneath. Curing onions should be checked regularly to avoid over-drying.

Emergency curing

If conditions such as rain or flooded fields do not permit field curing and curing facilities are not available, a temporary tent can be used for curing onions. In the example illustrated below, the tent is constructed from large tarps. Heated air is forced into a hollow area (known as a plenum) at the center of the bins of produce. Several fans are used to circulate the warm air through the onions while they are curing.

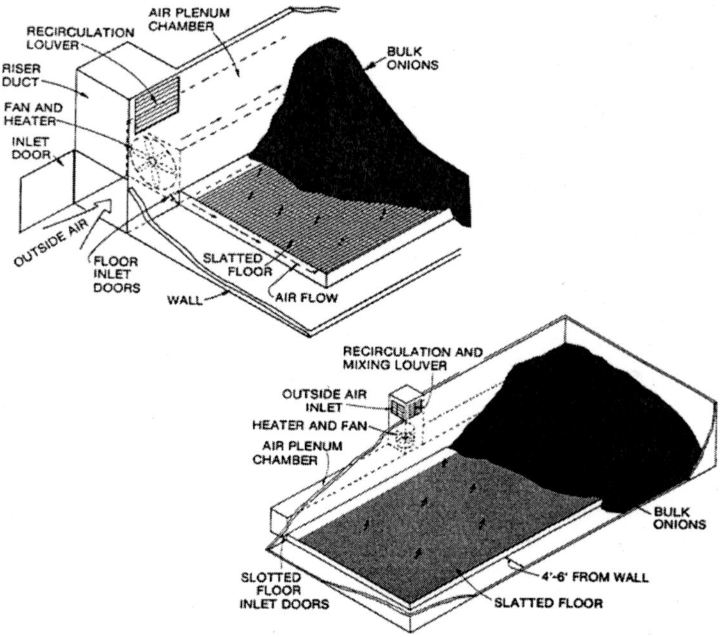

Figure 17. Bulk systems for curing onions

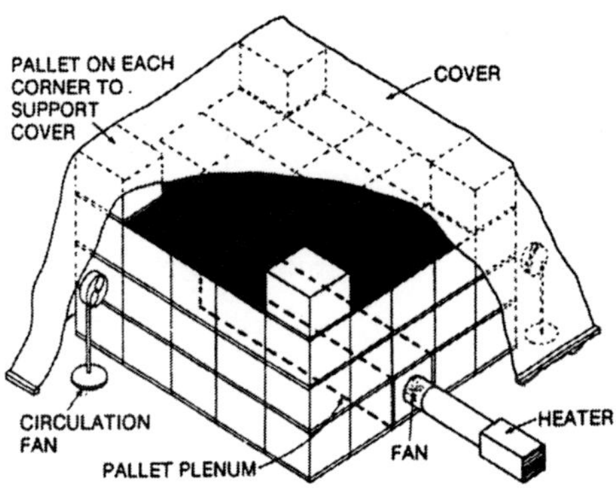

Figure 18. Emergency curing

DECAY AND INSECT CONTROL

The first line of defense against insects and disease is good management during production. Planting resistant varieties, the use of irrigation practices that do not wet the leaves or flowers of plants, avoiding over-fertilization with nitrogen, and pruning during production to reduce canopy overgrowth can all serve to reduce produce decay before and after harvest. A second important defense is careful harvesting and preparation for market in the field. Thirdly, sorting out damaged or decaying produce will limit contamination of the remaining, healthy produce. Yet, even when the greatest care is taken, sometimes produce must be treated to control insects or decay-causing organisms.

While high humidity in the storage environment is important for maintenance of high quality produce, any free water on the surface of commodities can enhance germination and penetration by pathogens. When cold commodities are removed from storage and left at higher ambient temperatures, moisture from the surrounding warm air condenses on the colder product's surfaces. A temporary increase in ventilation rate (using a fan) or increasing exposure of the commodity to drier air can help to evaporate the condensed moisture and to reduce the chances of infection.

Control of storage insects in nuts and dried fruits and vegetables can be achieved by freezing, cold storage (less than 5°C or 41°F), heat treatments, or the exclusion of oxygen (0.5% or lower) using nitrogen. Packaging in insect-proof containers is needed to prevent subsequent insect infestation.

Some plant materials are useful as natural pesticides. Cassava leaves are known to protect harvested cassava roots from pests when used as packing material in boxes or bags during transport and short-term storage. It is thought that the leaves release cyanogens, which are toxic to insects. The ashes of the leaves of *Lantana* spp. and *Ochroma logopur* have been found to be very effective when used as a dust against aphids attacking stored potatoes. The pesticidal properties of the seeds of the neem tree (as an oil or aqueous extract) are becoming more widely known and used throughout the world. Native to India , neem acts as a powerful pesticide on food crops but appears to be completely non-toxic to humans, mammals and beneficial insects. Any "natural pesticide" must be shown to be safe for humans before its approval by regulatory authorities.

Chemical Controls

Washing produce with chlorinated water can prevent decay caused by bacteria, mold and yeasts on the surface of produce. Calcium hypochlorite (powder) and sodium hypochlorite (liquid) are inexpensive and widely available. The effectiveness of the treatment will be decreased if organic matter is allowed to build up in the wash water.

The effectiveness of chlorine increases as pH is reduced from pH 11 to pH 8, but at lower pH chlorine becomes unstable.

Fruits and vegetables can be washed with hypochlorite solution (25 ppm available chlorine for two minutes) then rinsed to control bacterial decay. Alternatively, these commodities can be dipped in hypochlorite solution (50 to 70 ppm available chlorine) then rinsed with tap water for control of bacteria, yeasts and molds.

Sulfur: Sulfur is used on bananas as a paste (0.1 % active ingredient) to control crown rot fungi.

Sulfur dioxide: SO_2 is used as a fumigant (with a 10 ppm residue tolerance) on grapes to control Botrytis, Rhizopus and Aspergillus fungi. Careful calculation of the amount of sulfur dioxide required to treat grapes can greatly reduce the need to vent or scrub the storage air after fumigation to remove excess SO2. For information on the "total utilization" fumigation technique that has been developed for treating grapes with sulfur dioxide.

Sodium or potassium bisulfite: Bisulfites are usually contained within a paper or plastic pad that can be placed inside a carton to release SO_2 for control of molds on grapes during transportation and storage.

The use of bicarbonate salts for prevention of postharvest decay has been used on fresh peppers, melons, potatoes, carrots and citrus fruits. These salts are very inexpensive, safe to use, readily available and accepted as "certified organic" and "chemical free" for marketing purposes.

Applying lime powder (press butt-end into powder): On occasions when fungicides must be applied to produce, a simple tray with holes punched in the bottom can be used hold the commodity while it is sprayed.

A back-pack sprayer is a useful tool for applying fungicides and other postharvest chemical treatments. A wide variety of sizes and types of sprayers can be purchased at farm supply centers or via companies on the internet.

When fruit is packed for export, fungicides are often applied to meet the requirements of international quality standards and to reduce deterioration during transport. The "cascade applicator" was developed to apply fungicide uniformly and effectively by using a liquid curtain to drench the fruit.

Fruit in a perforated plastic tray is introduced on a roller conveyor belt (not shown) into the applicator. Inside a simple fan shaped deflector creates a curtain of liquid fungicide. The fruit passes under the curtain where it is drenched, then out of the applicator to drain on a tilted return tray. The tank holds up to 50 liters of fungicide solution, and a pump is mounted at the level of the tank outlet. A filter is fitted on the top of the tank to remove foreign matter from the return flow of fungicide from the applicator box and the return tray.

TEMPERATURE AND HUMIDITY CONTROL

Throughout the period between harvest and consumption, temperature control has been found to be the most important factor in maintaining product quality. Fruits, vegetables and cut flowers are living, respiring tissues separated from their parent plant. Keeping products at their lowest safe temperature will increase storage life by lowering respiration rate, decreasing sensitivity to ethylene gas and reducing water loss. Reducing the rate of water loss slows the rate of shriveling and wilting, causes of serious postharvest losses.

Keeping products too cool can also be a serious problem. It is important to avoid chilling injury, since symptoms include failure to ripen (bananas and tomatoes), development of pits or sunken areas (oranges, melons and cucumbers), brown discoloration (avocados, cherimoyas, eggplant), increased susceptibility to decay (cucumbers and beans), and development of off-flavors (tomatoes).

Cooling involves heat transfer from produce to a cooling medium such as a source of refrigeration. Heat transfer processes include conduction, convection, radiation and evaporation.

If a ready supply of electricity is available, mechanical refrigeration systems provide the most reliable source of cold. Methods include room cooling, forced-air cooling and evaporative cooling. A variety of portable forced-air coolers have been designed for use by small-scale growers and handlers. However, a variety of simple methods exist for cooling produce where electricity is unavailable or too expensive. Some examples of alternative systems include night air ventilation, radiant cooling, evaporative cooling, the use of ice and underground (root cellars, field clamps, caves) or high altitude storage. Ice can be manufactured using simple solar cooling systems, where flat plate solar collectors are used to generate power to make ice, which is then used to cool produce. Ice can be used either directly as package ice, to cool water for use in a hydro-cooler, or as an ice bank for a small forced air or room cooling system.

Several simple practices are useful for cooling and enhancing storage system efficiency wherever they are used, and especially in developing countries, where energy availability may be limited and any savings may be critical. Shade should be provided over harvested produce, packing areas, for buildings used for cooling and storage and for transport vehicles. Using shade wherever possible will help to reduce the temperatures of incoming produce and will reduce subsequent cooling costs. Trees are a fine source of shade and can reduce ambient temperatures around packinghouses and storage areas. Light colors on buildings will reflect light (and heat) and reduce heat load.

Sometimes spending money will save money, as when purchasing lighting equipment. High pressure sodium lights produce less heat and use less energy than incandescent bulbs.

Another aspect to consider when handling fruits and vegetables is the relative humidity of the storage environment. Loss of water from produce is often associated with a loss of quality, as visual changes such as wilting or shriveling and textural changes can take place. If using mechanical refrigeration for cooling, the larger the area of the refrigerator coils, the higher the relative humidity in the cold room will remain. It pays however, to remember that water loss may not always be undesirable, for example if produce is destined for dehydration or canning.

For fresh market produce, any method of increasing the relative humidity of the storage environment (or decreasing the vapor pressure deficit (VPD) between the commodity and its environment) will slow the rate of water loss. The best method of increasing relative humidity is to reduce temperature. Another method is to add moisture to the air around the commodity as mists, sprays, or, at last resort, by wetting the store room floor. Another way is to use vapor barriers such as waxes, polyethylene liners in boxes, coated boxes or a variety of inexpensive and recyclable packaging materials. Any added packaging materials will increase the difficulty of efficient cooling, so vented liners (about 5% of the total area of the liner) are recommended. The liner vents must line up with the package vents to facilitate cooling of the produce inside. Vented liners will decrease VPD without seriously interfering with oxygen, carbon dioxide and ethylene movement.

Room cooling

Room cooling is a relatively low cost but very slow method of cooling when electricity for mechanical refrigeration is available. When using room cooling, produce is simply loaded into a cold room, and cold air is allowed to circulate among the cartons, sacks, bins or bulk load. This cooling method is best suited to less perishable commodities such as potatoes, onions, apples, sweetpotatoes and citrus fruits, since more highly perishable crops will deteriorate too much before being adequately cooled. Room cooling may be all you need if you handle chilling sensitive crops that need to be cooled from early morning harvest temperatures to storage temperatures of 10 to 13 ºC (50-55 ºF). The design and operation of cold rooms are fairly simple and no special equipment is required.

It is important to leave adequate space between stacks of boxes inside the refrigerated room in order for produce to cool more quickly. About 1 inch (2.5 cm) is sufficient to allow cold air to circulate around individual boxes. Produce in vented boxes will cool much faster than produce packed in un-vented containers. In many small-scale cold rooms, produce has been loaded into the room so tightly that cooling cannot take place at all, and despite the high cost of running the refrigeration system, the produce temperature never decreases to recommended levels.

Stacks of produce inside the cold room should be narrow, about one pallet width in depth (two or three cartons). Fans should be installed to move the cold air throughout the room. Air circulating through the room passes over surfaces and through any open space, so cooling from the outside to the center of the stacks is mostly by conduction. You'll want to monitor the temperature of the produce within the packages at various locations in the room to determine that the produce is being cooled as desired. Rearrange the stacks and measure the rate of cooling until you find the right pattern for your cold room.

Forced-air cooling

Forced-air cooling pulls or pushes air through the storage containers themselves, greatly speeding the cooling rate of any type of produce. Many types of forced-air coolers can be designed to move cold moist air over the commodities. The example provided below is a fixed unit, where a fan is housed inside the wall of a cold room.

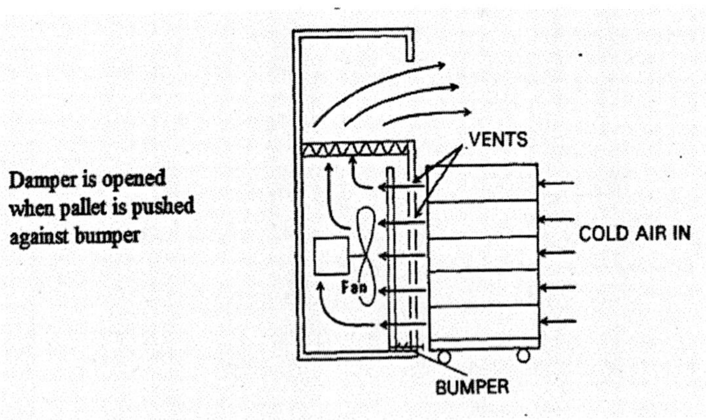

Figure 19. Cold wall forced-air cooler

A portable forced-air cooler can be constructed using a canvas or polyethylene sheet. The sheet is rolled over the top and down the back of the boxes to the floor, sealing off the unit and forcing air to be pulled through the vents (vent area should be at least 5% of the surface area of the carton) of the cartons stacked against the cooler. This unit is designed to be used inside a refrigerated storage room. The fan unit is shown detached to illustrate how the air should flow within the cooler. For best results and minimum cost of operation, the warm exhaust air from the fan should be directed toward the return air inside the cold room.

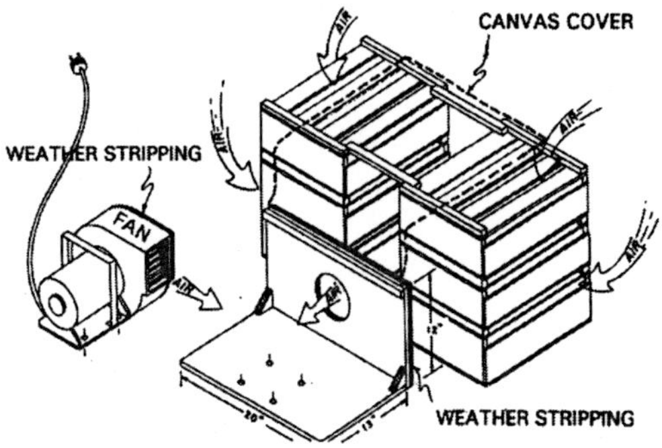

Figure 20. A portable forced-air cooler

Hydro-cooling

Hydro-cooling provides fast, uniform cooling for some commodities. The commodity as well its packaging materials must be tolerant of wetting, chlorine (used to sanitize the hydro-cooling water) and water beating damage. The simplest version of a hydro-cooler is a tank of cold water in which produce is immersed. The type shown below showers a batch of produce with icy water as the produce moves along a conveyor. A batch-type hydro-cooler can be constructed to hold entire pallet-loads of produce. Conveyors can be added to help control the time produce stays in contact with the cold water.

Evaporative Cooling

These packinghouses are made from natural materials that can be moistened with water. Wetting the walls and roof first thing in the morning creates conditions for evaporative cooling of a packinghouse that is made from straw.

Evaporative coolers can be constructed to cool the air in an entire storage structure or just a few containers of produce. These coolers are best suited to lower humidity regions, since the degree of cooling is limited to 1 to 2 C (2 to 4 F) above the wet-bulb temperature. A cooling pad of wood fiber or straw is moistened and air is pulled through the pad using a small fan. In the example provided here, O. 5 gallon of water per minute is dripped onto an 8 square foot pad, providing enough moist air to cool up to 18 crates of produce in 1 to 2 hours. Water is collected in a tray at the base of the unit and re-circulated.

An evaporative cooler can be combined with a forced air cooler for small lots of produce. Air is cooled by passing through the wet pad before it passes through the packages and around the produce. The air can be cooled to within a few degrees of the wet bulb temperature of ambient air.

Evaporative coolers can be constructed from simple materials, such as burlap and bamboo. The "drip cooler" shown here operates solely through the process of evaporation, without the use of a fan. Cooling will be enhanced if the unit is kept shaded and used in a well ventilated area.

An evaporative cooler located in the peak of a storage structure can cool an entire room of stored produce such as sweetpotatoes or other chilling sensitive crops. The vents for outside air should be located at the base of the building so that cool air is circulated throughout the room before it can exit.

Night air Ventilation

Storage structures can be cooled using night air if the difference in day and night temperature is relatively large. The storage facility should be well insulated and vents should be located at ground level. Vents can be opened at night, and fans can be used to pull cool air through the storeroom. The structure will best maintain cool temperatures during the heat of the day if it is well insulated and vents are closed early in the morning.

REFERNCES

Aiyer, RS. et al. 1978. "No-cost method for preserving fresh cassava roots". *Cassava Newsletter* 4: 8-9. Cali, Colombia: CIAT.

Bachmann, J. and R. Earles. 2000. "Postharvest Handling of Fruits and Vegetables". *ATTRA Horticulture Technical Note*. 19 pp.

Fellows, P. and Hampton, A 1992. *Small-Scale Food Processing A Guide to Appropriate Equipment*. London: Intermediate Technology Publications. 158 pp.

Harvey, E. et al. 1990. *Harvesting and postharvest handling of papayas in the Caribbean*. Bridgetown, Barbados: Inter-American Institute for Cooperation on Agriculture (IICA).

Lurie, S. 1998. "Postharvest heat treatments of horticultural crops". *Hort. Rev.* 22:91-121.

NRC. 1992. *Neem: A Tree for Solving Global Problems*. Washington, D.C.: Bostid Publishing Co. 141 pp.

Proctor, F.J. 1985. "Post-harvest handling of tropical fruit for export". *The Courier* 92: 83-86.

Shewfelt R.L. and Prussia, S.E. (eds). 1993. *Postharvest Handling A Systems Approach*. San Diego: Academic Press, 358 pp.

14

Packing of Horticultural Products

Packinghouse operations can be as simple as moving produce from a field lug into a shipping container, or may include a variety of handling practices, from cleaning, waxing, sizing, and quality grading to color sorting. The provision of shade during the packing operations is extremely important. Shade can be created using palm leaf fronds, a plastic mesh or canvas sheet hung from temporary poles, or via a permanent roofed structure. When deciding upon where to locate a packinghouse, access to the field and market point, adequate space for vehicles to enter and leave the packinghouse and ease of access to labor will all be considerations.

In the simplest packinghouse, produce is delivered in picking containers, immediately after harvest, directly to the packers. The packers then sort, grade, size and pack the produce directly into appropriate transport containers. In this case, each worker must be knowledgeable regarding produce defects, grade and size requirements, and packing methods. As the size and complexity of the packinghouse increases, more operations and workers trained in specific tasks might be added.

The typical series of operations in a packinghouse are illustrated in this chapter. Dumping can be done using either dry or water-assisted methods, depending upon the sort of produce being handled. Cleaning, as well, can be by washing with chlorinated water or dry brushing alone. Waxing, if practiced, occurs after washing and removal of surface moisture. Grading separates the product into processing and fresh market categories. Sizing further separates the product, with the smallest size going to the local market or to processing. Typically, the best quality produce is packaged and marketed at the regional or national level.

DUMPING

Produce must somehow be removed from the field bin or harvesting container and moved through the packinghouse. This first step is known as "dumping". Dumping must be done gently, whether using water assisted methods or dry dumping. Wet dumping can decrease bruising and abrasions by using moving, chlorinated (100-150 ppm) water to carry delicate produce. When using dry dumping, padded, sloped ramps or moving conveyor belts can decrease injuries to produce.

Any time produce is dumped from one container into another, care should be taken to reduce mechanical damage to the commodity. When dumping produce from field bins or from transport vehicles into the packinghouse, dry or wet dumping can be practiced. When using dry dumping practices, the filed container should be emptied slowly and gently onto a tilted ramp with padded edges. In the illustration below, a conveyor belt then carries the dry dumped produce into the packinghouse.

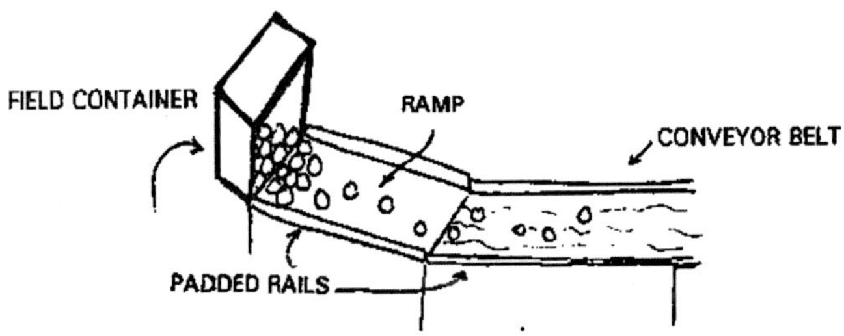

Figure 1. Dry dumping

Wet dumping is sometimes used to reduce mechanical damage, either by dumping into water rather than onto a dry ramp, or by immersion and floatation. If the specific density of the produce, such as apples, is lower than that of water the produce will float. For some produce, such as pears, salts (such as sodium lignin sulfonate, sodium silicate or sodium sulfate) must be added to the water to increase its specific density and assure fruit floatation. The canvas curtain illustrated below is used to break the fall of fruit moving from a conveyor into a bulk bin.

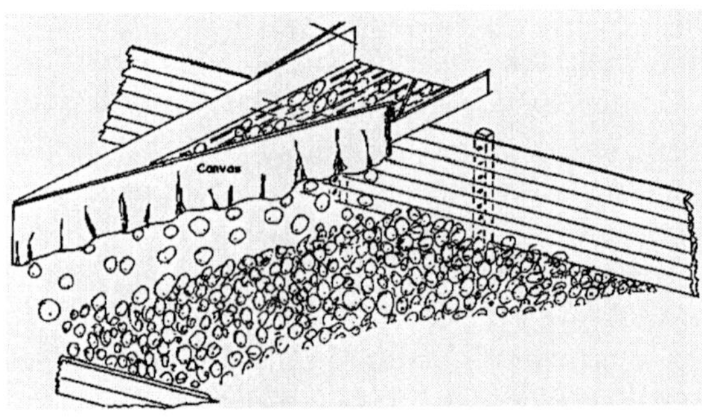

Figure 2. Canvas curtain

PRE-SORTING

Pre-sorting produce is usually done to eliminate injured, decayed, or otherwise defective produce (culls) before cooling or additional handling. Pre-sorting will save energy in that culls will not be handled. Removing decaying produce items will limit the spread of infection to other units, especially if postharvest pesticides are not being used.

CLEANING

For some commodities, such as kiwifruits and avocadoes, dry brushing may be sufficient to clean the produce. Other commodities, however, such as bananas and carrots, require washing. The choice of brushing and/or washing will depend upon both the type of commodity and the type of contamination.

Sanitation is essential, both to control the spread of disease from one item to another, and to limit spore buildup in wash water or in the packinghouse air. Chlorine treatments (100 to 150 ppm Cl) can be used in wash water to help control pathogen buildup during packing operations. There is some variation in the strength of bleach available commercially in different countries, but a rule of thumb is to use 1 to 2 mls of chlorine bleach per liter (1 to 2 ounces of chlorine bleach per 8 gallons of clean water). Walls, floors and packing equipment can also be cleaned using quarternary ammonium compounds labelled as safe for food processing equipment.

Washing

Steel drums can be used to make a simple washing stand. The drums are cut in halt fitted with drain holes and all the metal edges are covered with split rubber or plastic hose. The drums are then set into a sloped wooden table. The table top is constructed from wooden slats and is used as a drying rack before packing. Because steel drums are often used to store petroleum and chemical products, they should be thoroughly cleaned before being used as a washing stand. This tank for washing produce is made from galvanized sheet metal. A baffle made of perforated sheet metal is positioned near the drain pipe and helps to circulate water through the produce. Fresh water is added under pressure through a perforated pipe, which helps move floating produce toward the drain end of the tank for removal after cleaning.

WAXING

Waxing of immature fruit vegetables such as cucumbers and summer squash; mature fruit vegetables such as eggplant, peppers and tomatoes; and fruits such as apples and peaches is common. Food grade waxes are used to replace some of the natural waxes removed in washing and cleaning operations, and can help reduce water loss during handling and marketing. If produce is waxed, the wax coating must be allowed to dry thoroughly before further handling.

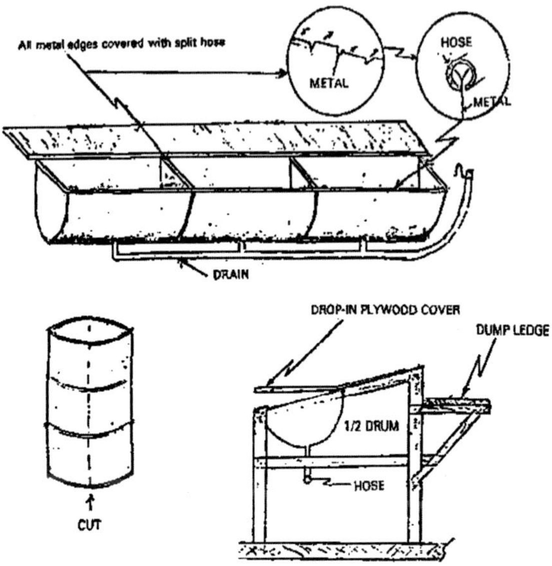

Figure 3. Washing stands

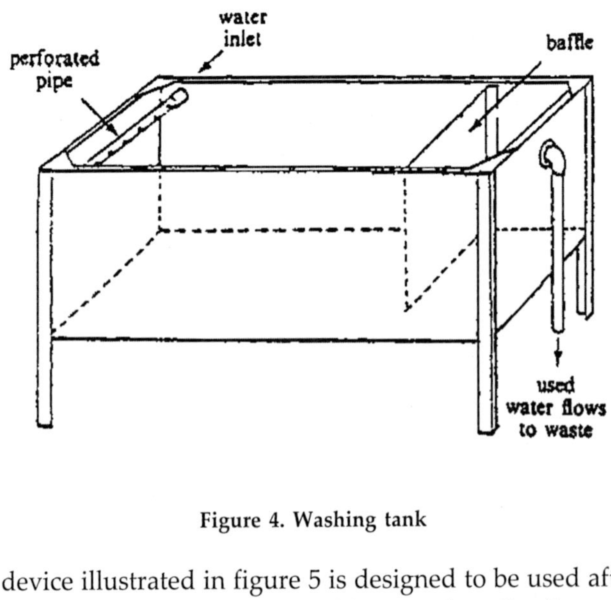

Figure 4. Washing tank

The waxing device illustrated in figure 5 is designed to be used after a series of dry brushes on a conveyor line. Industrial wool felt is used to distribute the liquid wax to the fruits or vegetables from a trough made the same width as the belt. Evaporation of wax from the felt is reduced by covering the felt with a layer of heavy polyethylene sheeting.

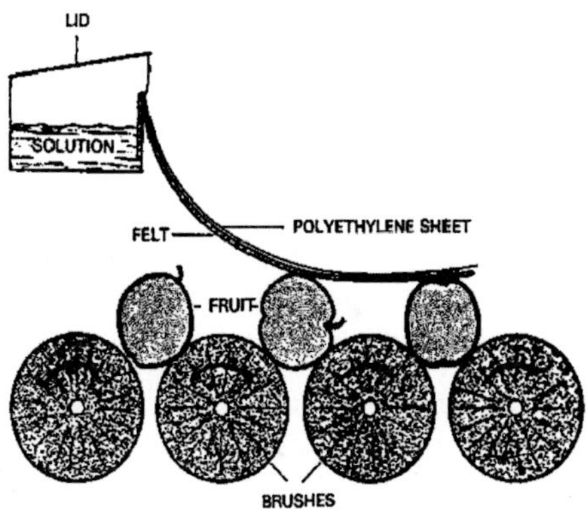

Figure 5. Waxing device

SORTING

The figure illustrated below is a combination sorting and packing stand. Incoming produce is placed in the sorting bin, sorted by one worker into the packing bin, and finally packed by a second worker. If workers must stand to sort produce, a firm rubber pad for the floor can help reduce fatigue.

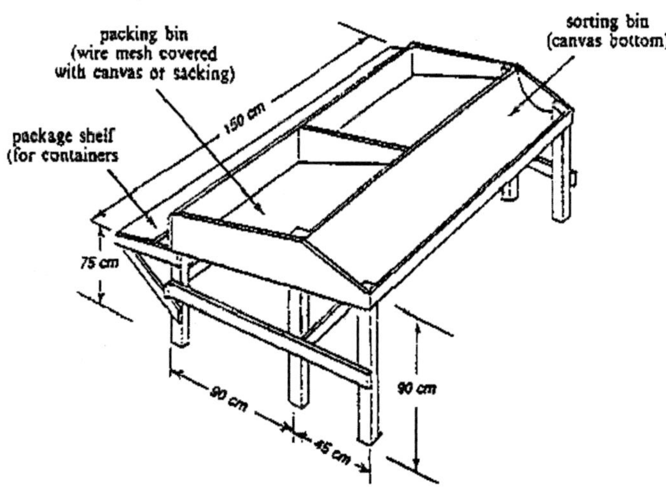

Figure 6. Sorting and packing stand

The surface of the portable sorting table illustrated below is constructed from canvas and has a radius of about 1 meter (about 3 feet). The edges are lined with a thin layer of foam to protect produce from bruising during sorting, and the slope from the center toward the sorter is set at 10 degrees. Produce can be dumped onto the table from a harvesting container, then sorted by size, color and/or grade, and packed directly into shipping containers. Up to 4 sorters/packers can work comfortably side by side.

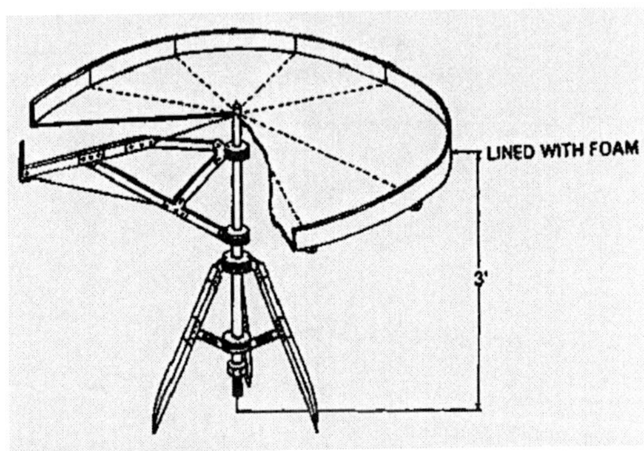

Figure 7. Portable sorting table

The following illustrations represent three types of conveyors used to aid sorting of produce. The simplest is a belt conveyor, where the sorter must handle the produce manually in order to see all sides and inspect for damage. A push-bar conveyor causes the produce to rotate forward as it is pushed past the sorters. A roller conveyor rotates the product backwards as it moves past the sorter.

Figure 8. Belt conveyor

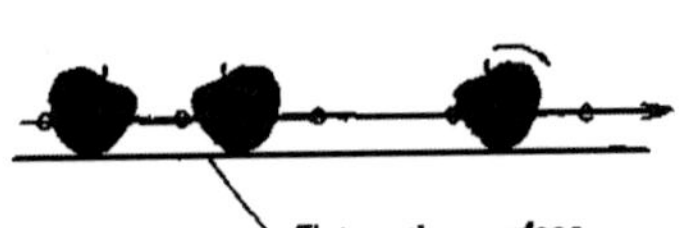

Figure 9. Push-bar conveyor

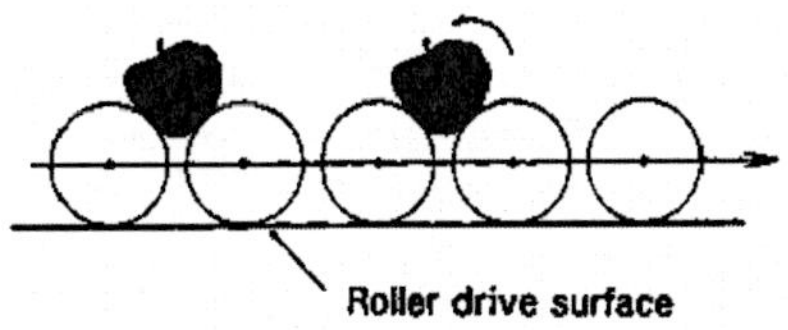

Figure 10. Roller conveyor

When sorting for rejects, and removing any product that is too small, decayed or damaged, the height of the sorting table should be set at a level comfortable for sorters. Stools, or a firm rubber pad on which to stand, can be provided to reduce fatigue. Locations of the table and the sorting bins should be chosen to minimize hand movements.

It is recommended that the workers' arms create a 45 degree angle when s/he reaches toward the table, and that the width of the table be less than 0.5 meter to reduce stretching. Good lighting will enhance the ability of the sorter to spot defects, and dark, dull belts or table tops can reduce eye strain.

If a conveyor system is in use, the product must not flow too fast for the sorters to do their work. The rotational speed of push-bar or roller conveyors should be adjusted to rotate the product twice within the immediate field of view of the worker.

SIZING

Sizing produce is optional but may be worthwhile if certain size grades receive a higher price than others. In most low-input packinghouses, manual sizing is still commonly practiced. Operators should be trained in selecting the size desired and to either directly pack the items into containers or place the selected produce gently into a bin for packing further down the line. Sizing can be done subjectively (visually) with the use of standard size guages. Examples of the smallest and largest acceptable sizes for each product can be placed within view of the operator for easy reference. Hand held sizers are used for a variety of products.

Several types of mechanical sizers are available for small scale operations. One type is composed of a long slanted tray with a series of openings which converge (largest at the top, smallest at the bottom). This type of sizer works best with round commodities. Other sizers are designed as conveyors fitted with chain or plastic belts with various sized openings, and are useful for sizing most commodities. Another simple method for mechanical sizing is to use a set of diverging bar rollers, where the smallest sized produce falls through the rollers first to a sorting belt or bin, and larger sized produce falls between successively more divergent rollers.

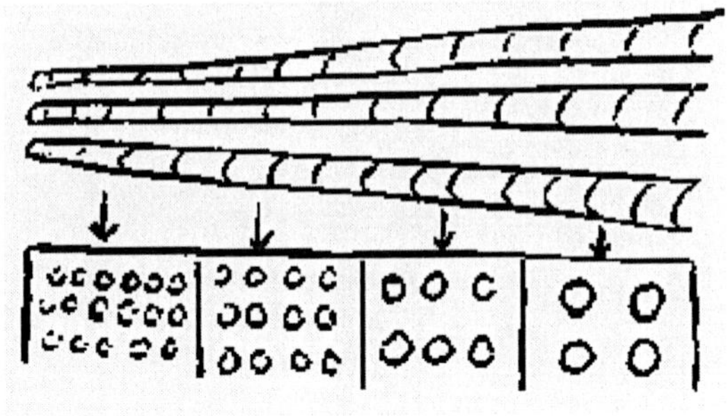

Figure 11. Diverging bar rollers

Round produce units can be graded by using sizing rings. Rings can be fashioned from wood or purchased ready-made in a wide variety of sizes.

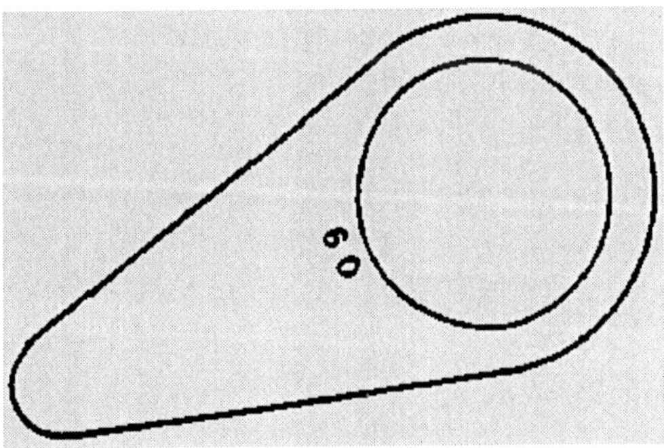

Figure 12. Single size hand held sizing ring

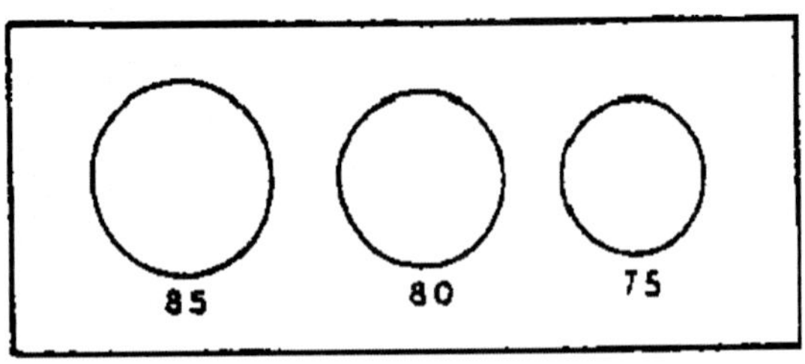

Figure 13. Multiple size rings

The rotary cylinder sizer illustrated below is composed of five hollow cylinders which rotate in a counterclockwise motion when driven by an electric motor. Each cylinder is perforated, with holes large enough to let fruits drop through. The first cylinder has the smallest diameter holes, and the fifth has the largest holes. When fruits fall through, they are caught on a slanted tray (the chute), and roll into the containers as shown. Take care that the distance of the drop is as short as possible to prevent bruising. Oversized fruits are accumulated at the end of the line. This equipment works best with round commodities.

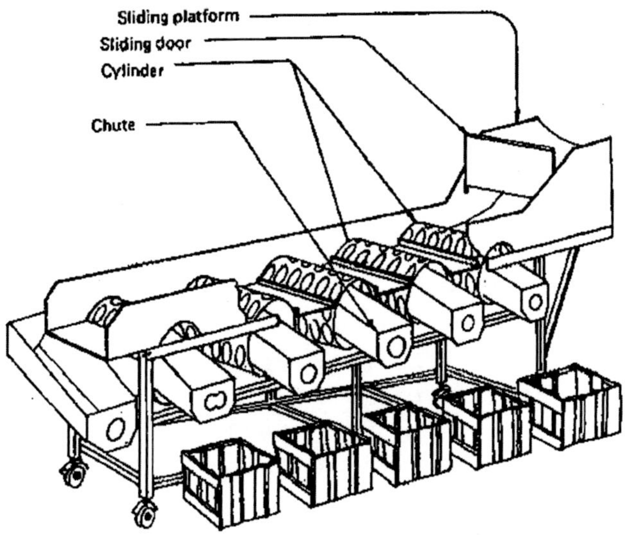

Figure 14. Rotary cylinder sizer

The onion sizing table illustrated below is one of three (or more) tables used in a stairway fashion. Each table is made of plywood, and has been perforated with holes of a specific size. The uppermost table has the largest size holes, and the lowest table has the smallest holes. A layer of onions is dumped onto the uppermost table. Those that do not pass through are classified as "extra-large" in size. Those that pass through fall into a mesh bag and roll into a large container. This container of onions is dumped onto the second sizing table. The onions that do not pass through are classified as "large", and so on.

The pommelo sizer illustrated below is composed of a rectangular chute made of plywood, padded with foam to prevent bruising. The fruit is dumped into the octagonal platform at the top of the chute, then allowed to roll, one by one, down toward a series of constrictions. Large fruits are caught in the first constriction, medium in the second, and small in the last. Undersized fruit passes out the end of the chute directly into a container. Workers must manually remove each fruit and place it into the appropriate size container before the next fruit can pass through the chute. The sizing is fastest when five workers are stationed at the sizer.

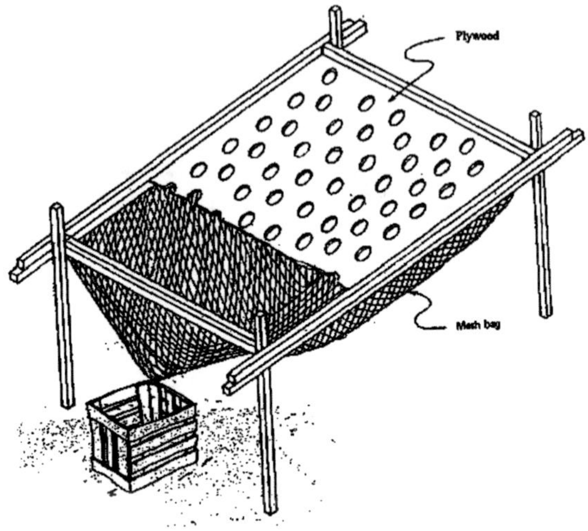

Figure 15. Onion sizing table

If a conveyor system is used in the packinghouse, a wide variety of sizing chains and belts are available for sorting produce. Sizing chains can be purchased in many widths and in any size opening.

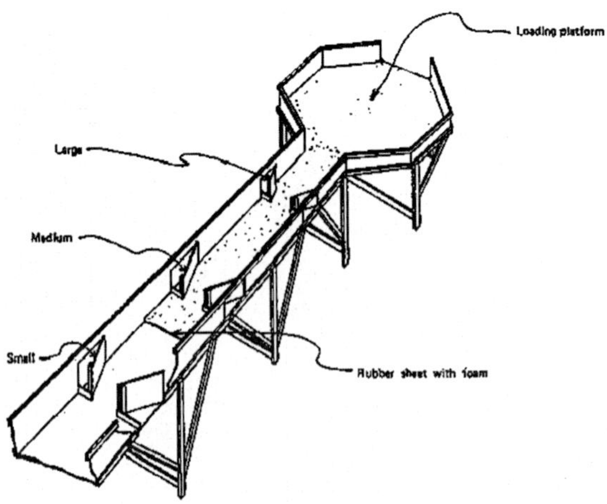

Figure 16. Pommelo sizer

Square openings are usually used for commodities such as apples, tomatoes and onions, while rectangular openings are used for peaches and peppers. Hexagonal openings are often used for potatoes and onions.

Small scale equipment for packing produce is available from several manufacturers and suppliers. Illustrated below is a fruit packing line, available from TEW Manufacturing Corporation at a cost of less than US$ 5000. This particular model includes a receiving belt, washer and sorting table.

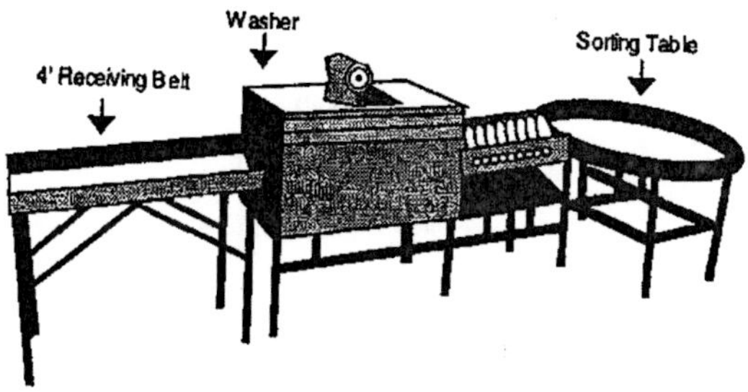

Figure 17. Fruit packing line

PACKING PRACTICES

If produce is packed for ease of handling, heavily waxed cartons, wooden crates or rigid plastic containers are preferable to bags or open baskets, since bags and baskets provide no protection to the produce when stacked. Sometimes locally constructed containers can be strengthened or lined to provide added protection to produce. Waxed cartons, wooden crates and plastic containers, while more expensive, are reusable and can stand up to the high relative humidity found in the storage environment. Containers should not be filled either too loosely or too tightly for best results. Loose products may vibrate against others and cause bruising, while overpacking results in compression bruising. Shredded newspaper is an inexpensive and lightweight filler for shipping containers.

For small-scale handlers interested in constructing their own cartons from corrugated fibreboard, Broustead and New provide detailed information. Many types of agricultural fibres are suitable for paper making, and handlers may find it economically sensible to include these operations in their postharvest system. Throughout the entire handling system, packaging can be both an aid and a hindrance to obtaining maximum storage life and quality. Packages need to be vented yet be sturdy enough to prevent collapse. Collapsed packages provide tattle or no protection, requiring the commodity inside to support all of the weight of the overhead load. Packing is meant to protect the commodity by immobilizing and cushioning it, but temperature management can be made more difficult if packing materials block ventilation holes. Packing materials can act as vapor barriers and can help maintain higher relative humidities within the package. In addition to protection, packaging allows quick handling throughout distribution and marketing and can minimize impacts of rough handling.

Produce can be hand-packed to create an attractive pack, often using a fixed count of uniformly sized units. Packaging materials such as trays, cups, wraps, liners and pads may be added to help immobilize the produce. Simple mechanical packing systems often use the volume-fill method or tight-fill method, in which sorted produce is delivered into boxes, then vibration settled. Most volume-fillers are designed to use weight as an estimate of volume, and final adjustments are done by hand.

Packaging in plastic films can modify the atmosphere surrounding the produce (modified atmosphere packaging or MAP). MAP generally restricts air movement, allowing the product's normal respiration processes to reduce oxygen content and increase carbon dioxide content of the air inside the package. An additional major benefit to the use of plastic films is the reduction of water loss.

MAP can be used within a shipping container and within consumer units. Atmospheric modification can be actively generated by creating a slight vacuum in a vapor sealed package (such as an unvented polyethylene bag), and then replacing the package atmosphere with the desired gas mixture. In general, lowering oxygen and increasing carbon dioxide concentrations will be beneficial for most commodities (see the table of recommended gas mixtures for various crops, Section 7). Selection of the best polymeric film for each commodity/package size combination depends upon film permeability and the respiration rate of the commodity under the expected time/ temperature conditions during handling. Absorbers of oxygen, carbon dioxide and/or ethylene can be used within packages or containers to help maintain the desired atmospheric composition.

Modified atmosphere packaging should always be considered as a supplement to proper temperature and relative humidity management. The differences between beneficial and harmful concentrations of oxygen and carbon dioxide for each kind of produce are relatively small, so great care must be taken when using these technologies.

The packing stand illustrated in the diagram below can be bolted to a second stand of the same construction if more space is required for packing produce. When trimming is necessary, add a loose board, thick enough to reach above the height of the front rail. The front rail should be smooth and rounded.

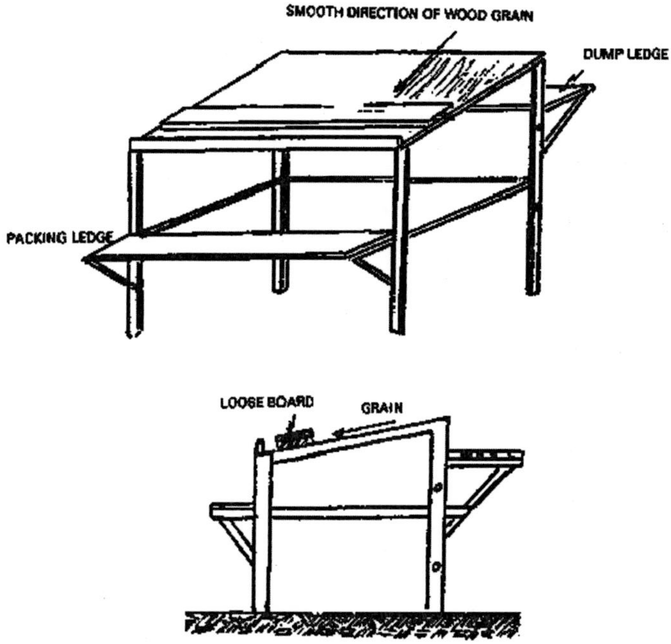

Figure 18. Packing stand

A simple field packing station can be constructed from wooden poles and a sheet of polyethylene. Thatch over the roof will provide shade and keep the station cool. The structure should be oriented so that the roof overhang keeps out the majority of the sun's rays.

Hands of bananas, after undergoing washing to remove latex and perhaps spraying with fungicides, are typically packed into cardboard containers lined with polyethylene. The following illustrations depict one method of filling a container with the fruit in order to ensure less damage during transport. Note that the polyethylene liner is folded up over the bananas before closing the box.

Figure 19(a). Wide, flat medium-to-small hand in middle of compartment

Figure 19 (b). Medium-length, wide hand on top, crown not touching fruit below

Figure 19 (c) Medium-to-short length, wide hand, crown not touching fruit below

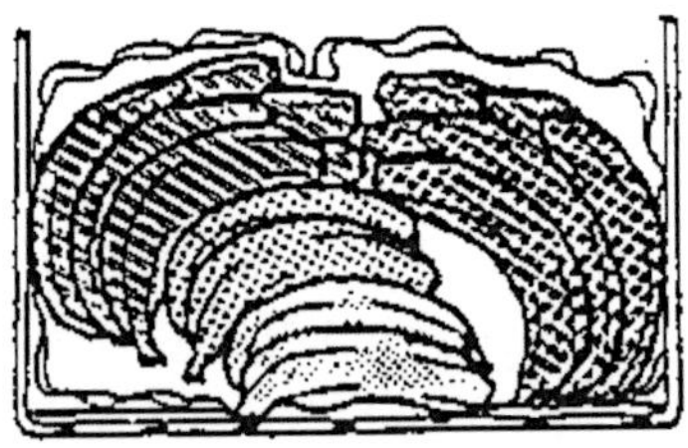

Figure 19 (d). One large hand, or two clusters with long fingers

A circular rotating table can be used to pack a variety of crops. The produce is fed in along a conveyor or if no conveyor is in use, simply put onto the table, where packers select the produce and fill cartons at their stations. In the illustration below, a discard belt has been added below the supply belt, allowing easy disposal of culls. Each packer can work independently, trimming as needed and check weighing cartons on occasion.

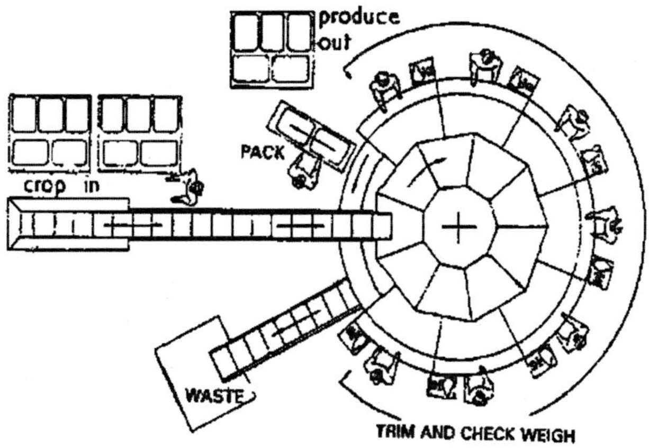

Figure 20. Circular rotating table

PACKING CONTAINERS

There are many types of packing containers. The three containers illustrated below are constructed from corrugated cardboard. The regular slotted container is fully collapsible and the most economical.

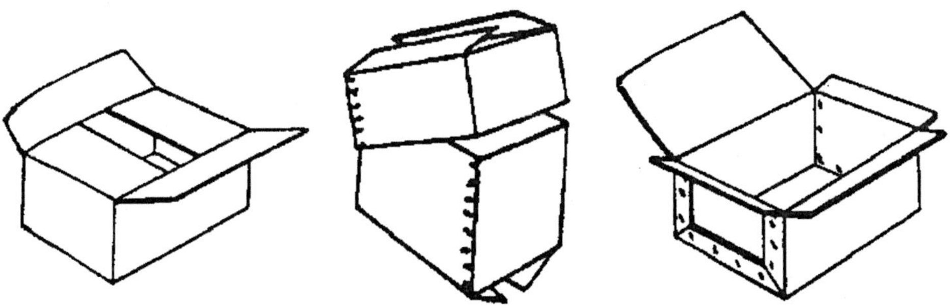

(a) (b) (c)

Figure 21. (a) Telescopic container; (b) Bliss box (c) Sack

Telescopic containers (half or full) have the highest stacking strength and protect against bulging but are more costly. The container known as a Bliss box has very strong corners, but is not collapsible. Sacks are often used to package produce, since they tend to be inexpensive and readily available.

The diagrams below are for a variety of commonly used fibreboard containers. Final dimensions can be altered to suit the needs of the handler.

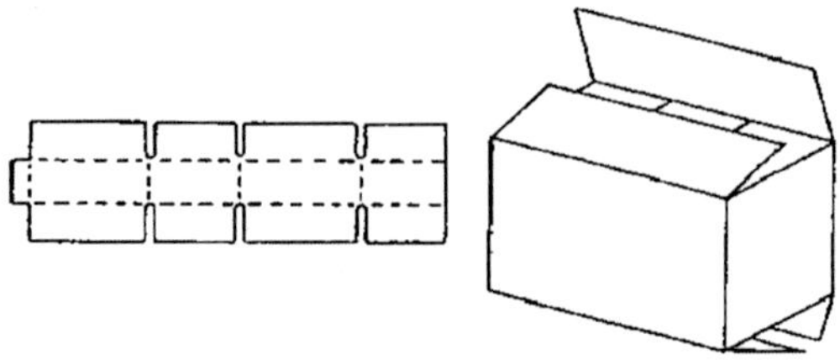

Figure 22. One piece box

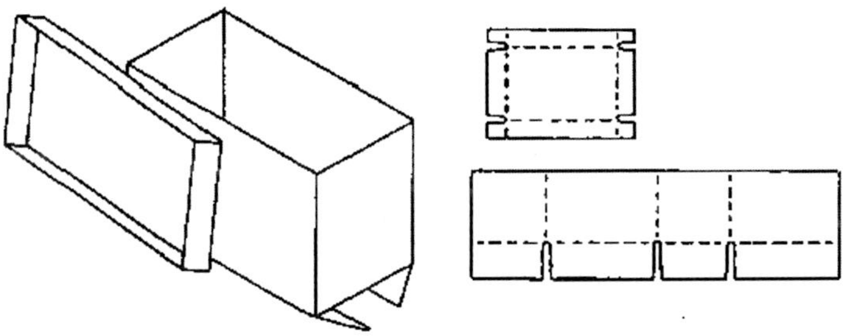

Figure 23. Two-piece box with cover

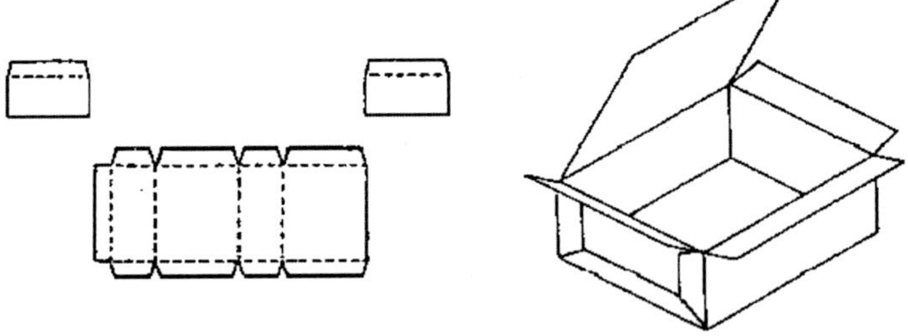

Figure 24. Bliss-style box

The diagrams below are for a variety of commonly used fibreboard containers. Final dimensions can be altered to suit the needs of the handler.

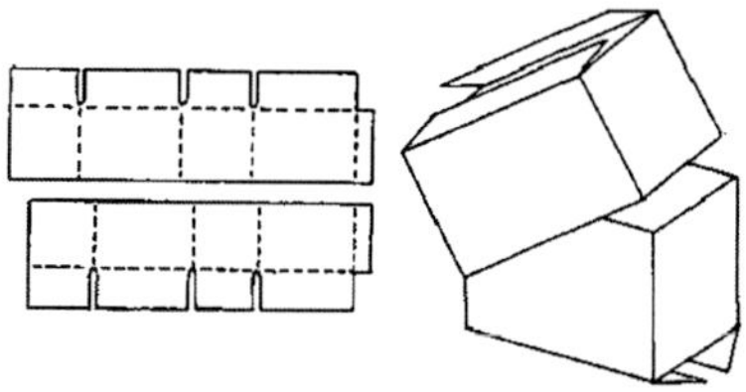

Figure 25. Full telescoping box

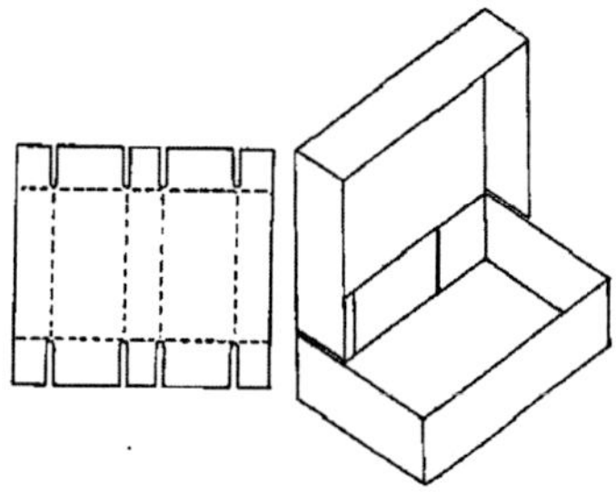

Figure 26. One-piece telescoping box

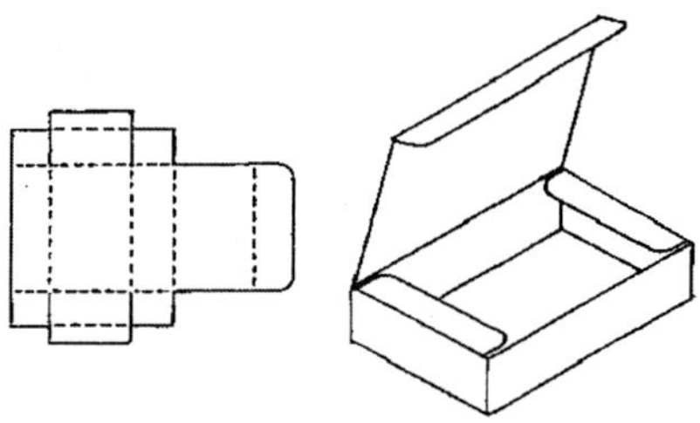

Figure 27. One-piece tuck-in cover box

The diagrams below are for a variety of commonly used fibreboard containers. Final dimensions can be altered to suit the needs of the handler.

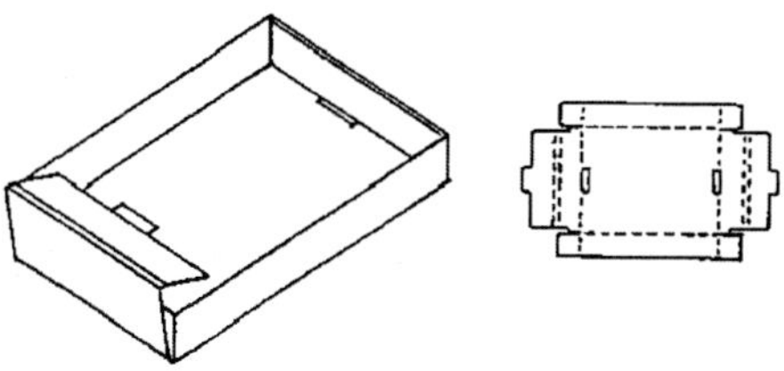

Figure 28. Self-locking tray

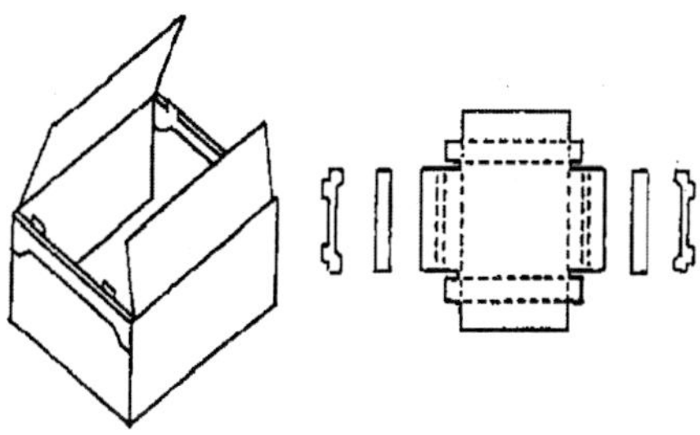

Figure 29. Interlocking box

Shipping containers can be designed and made by the user from fibreboard in any size and shape desired. Three types of joints are commonly used to construct sturdy boxes.

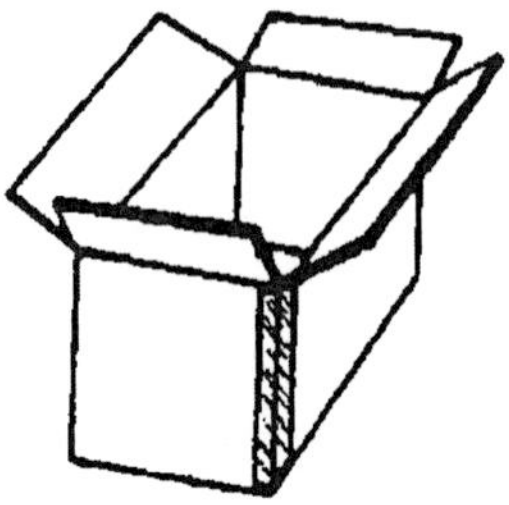

Figure 30. Taped joints

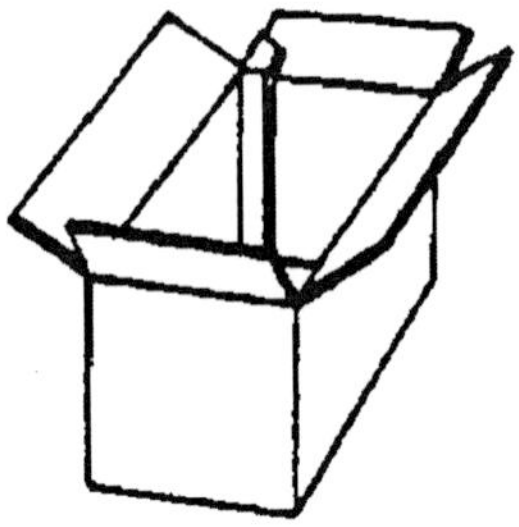

Figure 31. Glued Joints

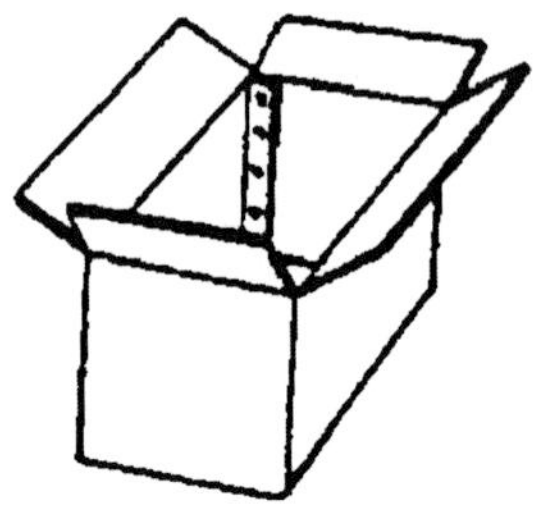

Figure 32. Stapled joints

Containers can be constructed from wood and wire. A special closing tool makes bending the wire loops on the crate's lid easier for packers to do. Wirebound crates are used for many commodities including melons, beans, eggplant, greens, peppers, squash and citrus fruits. Package Research Laboratory can provide a list of suppliers in your area.

A wooden lug is the typical packing container for table grapes. This container is very sturdy and maintains its stacking strength over long periods of time at high relative humidity. Rigid plastic containers are also widely used.

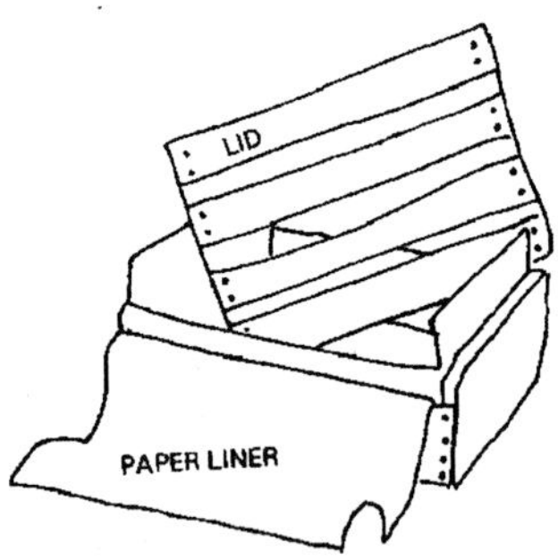

Figure 33. Wooden lug

Often, a paper liner is folded over the grapes before the top is nailed closed. The liner protects the produce from dust and water condensation. If a pad containing sulfur dioxide can be enclosed with the grapes within a plastic liner as a treatment to control decay. Most commodities other than table grapes can be damaged (bleached) by sulfur dioxide treatments.

Rigid plastic or wooden containers are also used extensively for asparagus. The trimmed spears are packed upright in containers that provide for a large amount of ventilation.

Containers for cut flowers are often long and narrow, of full telescopic design with vents at both ends to facilitate forced-air cooling. The total vent area should be 5% of the total box surface area. A closable flap can help maintain cool temperatures if boxes are temporarily delayed in transport or storage in an uncontrolled temperature environment. A simple wooden tray with raised corners is stackable and allows plenty of ventilation for fragile crops such as ripe tomatoes.

PACKAGING PRACTICES

Adding a fiberboard divider to a carton will increase stacking strength. The use of dividers is common with heavy crops such as melons. The dividers also prevent melons from vibrating against one another during handling and transport. Wooden inserts, or fiberboard folded into triangles and placed in all four corners can be especially useful when a carton needs strengthening.

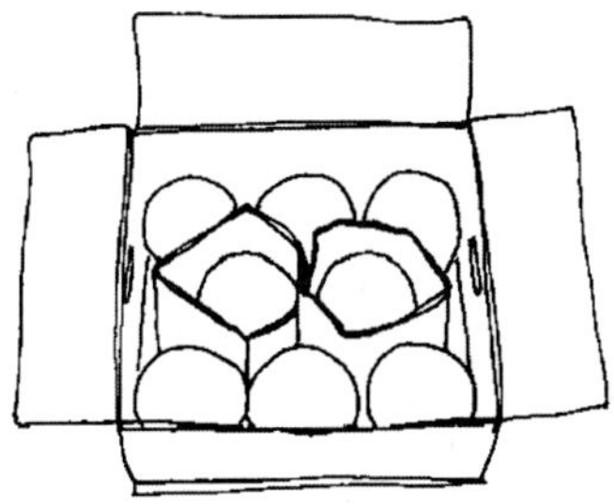

Figure 34. Fiberboard divider

When locally made containers have sharp edges or rough inner surfaces, a simple, inexpensive inner made from fiberboard can be used to protect produce from damage during handling.

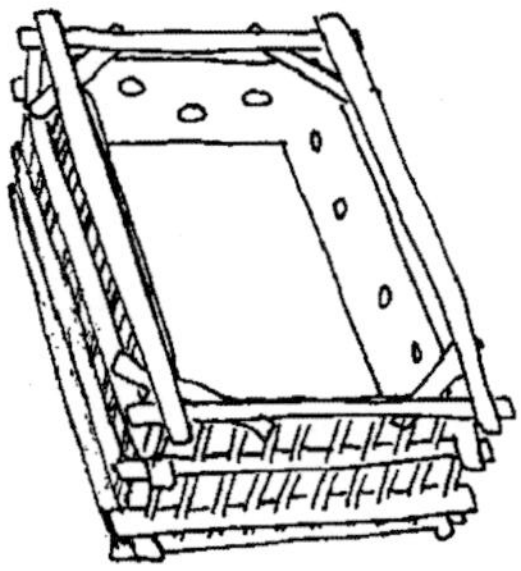

Figure 35. Cardboard liner for a palm rib crate:

If large bags or baskets must be used for bulk packaging of fruits or vegetables, the use of a simple vent can help reduce the buildup of heat as the product respires. Thin paper or plastic sleeves are a useful material for protecting cut flowers from damage during handling and transport.

LABELING

Labeling packages helps handlers to keep track of the produce as it moves through the postharvest system, and assists wholesalers and retailers in using proper practices. Labels can be preprinted on fiberboard boxes, or glued, stamped or stenciled on to containers. Brand labeling packages can aid in advertising for the product's producer, packer and/or shippers. Some shippers also provide brochures detailing storage methods or recipes for consumers.

Shipping labels can contain some or all of the following information:

- Common name of the product.
- Net weight, count and/or volume.
- Brand name.
- Name and address of packer or shipper.
- Country or region of origin.
- Size and grade.
- Recommended storage temperature.
- Special handling instructions.
- Names of approved waxes and/or pesticides used on the product.

Labeling of consumer packages is mandatory under FDA regulations. Labels must contain the name of the product, net weight, and name and address of the producer, packer or distributor.

MODULARIZATION OF CONTAINERS

When a variety of different sized cartons are packed at the same time, using boxes in standard sizes can greatly ease future handling. When handling boxes that are non-uniform, stacks can be unstable or heavier cartons can crush lighter ones. An unstable load is likely to fall over during transport or to collapse during storage. Recommended container sizes are shown below. These containers are part of the MUM program (Modularization, Unitization and Metrication) advocated by the USDA. They can all be stacked in a variety of patterns, depending upon their size, yet still form a stable load on a single pallet of 1000 x 1200 mm (40 x 48 inches).

Table 1. MUM containers for horticultural crops

mm	*Outside Dimensions* *Inches*	*Number Per Layer*	*Pallet Surface Area Utilized* *Percentage*
600 x 500	(23.62 x 19.69)	4	100
500 x 400	(19.68 x 15.75)	6	100
600 x 400	(23.62 x 15.75)	5	100
500 x 333	(19.68 x 13.11)	7	97
600 x 333	(23.62 x 13.11)	6	99
500 x 300	(19.68 x 11.81)	8	100
475 x 250	(18.70 x 9.84)	10	99
400 x 300	(15.75 x 11.81)	10	100
433 x 333	(17.01 x 13.11)	8	96
400 x 250	(15.74 x 9.84)	12	100

Figure 36. A pallet load of MUM containers

MODIFIED ATMOSPHERE PACKAGING (MAP)

Within a consumer package: If commodity and film permeability characteristics are properly matched, an appropriate atmosphere can evolve passively through consumption of O_2 and production of CO_2 during respiration. Some rigid plastic consumer packages are designed with a gas diffusion window.

Lightly processed lettuce (shredded or chopped) can be packaged in 5-mil plastic bags. After a partial vacuum is created, a gas mixture of 30 to 50% O_2 and 4 to 6% CO_2 is introduced into the bag, which is then sealed.

Within a shipping container: Polyethylene liners are added to shipping containers in cherry boxes, and polyethylene bags are used for bananas destined for distant markets.

Within a pallet: A single pallet load of produce such as strawberries can be sealed within a shroud of 5 mil polyethylene bag and a plastic sheet on the pallet base using wide tape. A slight vacuum can be introduced and 15% CO_2 added to the air introduced via a small hose.

Many plastic films are available for packaging, but very few have gas permeabilities that make them suitable for MAP. Low density polyethylene and polyvinyl chloride are the main films used in packaging fresh fruits and vegetables. Saran and polyester have such low gas permeabilities that they are suitable only for commodities with very low respiration rates.

UNIT LOADS

Many shippers and receivers prefer to handle unit loads of produce pallets rather than handling individual shipping containers. The switch to unit loads has reduced handling, causes less damage to the containers and produce inside, and allows faster loading/unloading of transportation vehicles. If small scale handlers wish to use unit loads for shipping produce, either wooden pallets or slip sheets can serve as the base of the load. Using guides for aligning the boxes (such as placing the pallet to be loaded against the corner of a room, or building a set of "bounce boards" if the pallet is loaded outside) will stabilize the load. Using fiberboard, plastic or wooden containers with vertical interlocking tabs can also help improve the unit load's stability. Containers must have holes for ventilation which align when stacked squarely on top of one another. Glue can be used between layers of containers to reduce slipping, and plastic netting or plastic or metal straps should be added to secure the load. Cornerboards made from fiberboard, plastic or metal help to provide for a stable unit load.

REFERENCES

Thompson, A.K. 1996. *Postharvest Technology of Fruits and Vegetables.* Blackwell Science, Ltd., Oxford, UK, 410 pp.

Wills, R., B. McGlasson, D. Graham, and D. Joyce. 1998. *Postharvest: An Introduction to the Physiology and Handling of Fruit, Vegetables, and Ornamentals.* CAB International, Wallingford, UK, 262 pp.

Zagory, D. 1999. "Effects of post-processing handling and packaging on microbial populations". *Postharv. Biol. Technol.* 15:313-321.

15

Horticultural Marketing

Marketing is the process by which the space between the producer and the consumer is bridged. The process obviously involves transport and techniques for minimising crop losses. An effective distribution system will also require the establishment of rural businesses such as truck drivers and packaging manufacturers, contractors and wholesalers.

The production/marketing chain is a twoway process. Produce flows from the rural areas into the cities and money and market information should flow back. As tastes in the city market evolve the rural community can use this market information to target its production accordingly.

In horticultural farming, where prices are rarely regulated, financial viability depends as much upon business and marketing skills as on the farmer's technical expertise. It is high-value crops which are often a crucial component of viable small farms. This manual is a response to that growing farmer need for commercial and marketing knowledge and is an entirely practical handbook. The techniques and advice have been tested and proven in the field.

Marketing involves finding out what your customers want and supplying it to them at a profit. This description stresses the two crucial points that govern marketing:

- Firstly, that the whole marketing process has to be customer oriented. Production must supply customers with what they want or need. This is the only reason people spend their money.
- Secondly, that marketing is a commercial process and is only sustainable if it provides all the participants with a profit.

A classical definition is: The series of services involved in moving a product from the point of production to the point of consumption.

This definition emphasizes that agricultural marketing is achieved by a series of processes. In this we include harvesting techniques, the grading and sorting of crops and the packing, transport, storage, processing, distribution and selling of products. These are the mechanics of marketing.

A broader view of marketing is provided by the following definition:

> The series of activities involved in making available services and information which influence the desired level of production relative to market requirements, and the movement of the product from the point of production to the point of consumption.

This definition covers the services which should be covered by the extension officer, such as providing information and advice. This role includes:

- collection, evaluation and dissemination of market information;
- assistance in the planning and scheduling of production;
- securing the market for producers, e.g. through contracts with buyers;
- advice on the best practical post-harvest practices;
- coordination of inputs, transport, storage, credit and post-harvest facilities.

Although this definition is more comprehensive it still only describes the activities involved in marketing. The key activity of an extension officer, or any public servant concerned with improving agricultural marketing, is the commercialisation of the rural economy. This involves: finding out what the customer wants and helping to set up the production/marketing system which supplies that demand and maximises the income of rural areas. Progress is achieved by concerted effort. Very often the marketing officer's role is to take an overview and coordinate the efforts of others.

IMPORTANCE OF GOOD MARKETING

The importance of good marketing can be conveniently considered from the four different perspectives of the national economy, the farmer, the product and the consumer. At the national level as societies and countries develop there is a movement of people from the countryside into the towns and cities.

Populations in developing countries are expanding, normally at around three percent a year. Urban populations, however, are expanding on average at about four percent a year. This means that the number of people needing to be fed by the rural communities will double in 16 years. In addition, since the amount of food eaten by each individual normally increases as people become wealthier, the supply of food for the towns and cities will need to double approximately every 10 to 14 years.This change in population distribution will create new or improved opportunities for both farmers and rural employment, particularly if new roads and improved transport are provided.

Subsistence farming will become less important. Although proportionately there will be fewer farmers, their role will become ever more important because their task will be to feed the growing urban populations. To do so will require the farmer to become more specialised and more skilled so that more food can be produced.

The extension officer's first role in marketing is to guide and assist farmers in the process of change from subsistence farming to commercial farming. There is a role for him or her at each stage in the development of agriculture, encouraging farmers to develop new skills needed to market and sell their produce. Even in highly developed societies where farmers are acknowledged as highly skilled producers they are very often weakest at marketing. The second important role is to try to secure and maximise rural incomes.

The reasons why people move from the countryside into the town vary from country to country and from case to case. Possibly the most important factor is the high relative incomes that can be earned in the towns and on regular employment. An important part of the extension officer's role should be to encourage and help the rural community to take control and develop the marketing of its food products. They should try to ensure that the maximum proportion of the retail price is circulated back to the rural areas.

At the farmer level the most disadvantaged farmers are those with small units of land. These farmers will find that they cannot generate sufficient funds from their small land area to support themselves and their families by growing only traditional crops, e.g. wheat and rice. They will find it difficult to compete with produce grown by large farms using mechanisation.

In developed agriculture successful small farms can and do survive. We can learn lessons from their survival. Viable small farms tend to specialise in highoutput enterprises. These are crop or livestock systems which are capable of generating high incomes per unit of land. Typical examples in livestock are dairy and chickens, and in crops are fruit, vegetables and flowers.

It is by understanding the strengths and weaknesses of both groups of farmers that it is possible to promote crops and cropping systems which favour the smaller farm. These growers need help in access to markets, by being provided with good production advice and market information to strengthen their ability to negotiate.

HORTICULTURAL MAREKTING PROCESS

Horticultural products are mainly sold fresh; some are eaten raw while others are cooked. Some horticultural products have traditionally been processed when no other form of storage was available, e.g. dried fruit and jams. As society develops and becomes more affluent the market for processed and prepared horticultural products develops. A market also develops for horticultural products such as flowers and house and garden plants which are sold for purely aesthetic reasons. Increased wealth also brings with it an increased demand for product diversity in the form of new crops, off-season supplies and different flavours

Horticultural products are perhaps most easily defined as what they are not. They are not cereals or the major industrial crops. Generally, but not exclusively, they are not staple crops.

Important characteristics of horticultural crops are:

- that they are mainly eaten for their contribution to the flavour and interest of food and for the supply of minor but essential nutrients, especially vitamins;
- that they are not basic food commodities; people will put off buying if the price is too high;
- that consumption levels vary, depending on the selling price and the income of the buyer,
- that many of the crops are not traded in large volumes and there is a limited market;
- that the products are perishable, which means there is always a reduction in quality if they are not sold immediately, usually leading to a fall in value;
- that there is a wide range and variety of horticultural products. If one product is too highly priced the consumer will generally buy another;
- that the products are normally traded in a very free market where price is determined by supply and demand.

All these factors contribute to the crucial and reoccurring fact about horticultural crops: that prices, and especially the prices the farmers obtain, are variable and difficult to predict.

Take an extreme example, but one that occurs regularly:

- A farmer who has high quality tomatoes to sell when few other crops are available may easily get a price equivalent to many times his growing cost. However, a farmer who is trying to sell tomatoes when the market is oversupplied and his fruit is two days old may not be able to sell his produce at all.
- Wholesale prices may double or halve in the same day, depending on the skill of the salespeople and on consumer demand.

Horticultural crop prices can fluctuate widely:

- from year to year;

- from the start of the season to the main supply period;
- from day to day; and
- from market to market.

This extreme variance in price makes horticultural production potentially both very profitable and very risky. Often, success depends on marketing skills and on obtaining good prices rather than on production expertise. The marketing/asks of an extension officer can be conveniently divided into three stages.

The first stage is research and analysis. Put more simply this means finding out the problems and opportunities of horticultural growers in the area. It involves talking to growers and traders, finding out prices, understanding how the marketing system works and investigating how the farmers can increase their income by selling crops.

The second stage is deciding what to do and, more importantly, getting agreement from those involved on the best course of action. Obviously, correct decisions are dependent on how thoroughly the first stage has been carried out.

The third stage is converting the plan into action. This stage will involve not only providing marketing advice to farmers, but also to the businesses in the production/ marketing chain.

The research and analysis phase of the work can itself be easily divided into three steps.

The first involves carrying out an investigation of the area/region/country. The work will need to cover everything from crop production to transport links to the market. The extension officer should attempt to find comparative advantages in his area and identify problems which need to be overcome.

The second step is to find out from the market what product or products are wanted and in what form. The work involves market research and is concerned with finding out the customers' requirements.

The third step covers understanding how the marketing system works, who is involved and how it wants to be serviced.

Audit of Local Resources and Facilities

The objective is for the extension officer to thoroughly familiarise him or herself with both the problems as well as the opportunities of the area. Set out below is a checklist of most of the important questions that should be answered.

It is very important to speak to farmers. In particular the extension officers should ensure they meet average and small farmers. Generally farmers have a good understanding of their problems and are delighted to have an opportunity to talk. The extension officer's role is to listen and learn. He or she should try to understand how farmers might react to new ideas and which farmers are likely to be most positive. The extension officer should find out where farmers meet to discuss matters and whose opinions they particularly respect.

At the end of this stage the extension officer should have a clear idea of the crops, the marketing system, the individuals and the problems of the area. He or she will also have some idea of some possible solutions which are worth investigating. The break-even price of delivering produce to the most likely markets should have been calculated.

Market Wants in Terms of Product

By customer we refer to the individuals in the marketing chain who buy and sell the produce, as well as the final consumer. We need to know who currently supplies the market, at what times and at what prices? Also, what volumes are sold and how the produce is packed and presented. As the objective in most eases will be either to start or to increase the supply of produce the extension officer should try to understand what the effect of this will be on the market. It could, for example, result in oversupply which will force down prices or it may supply a previously unsatisfied demand, which of course is the ideal situation. The critical questions are: what volume can sold; during which periods is it best to supply the market; and how should the produce be graded and packed?

Building up answers to these questions is again a process of information gathering. The sources are varied and certainly no single source is satisfactory. The extension officer should seek opinions from knowledgeable individuals, particularly those who are commercially involved with trading. In addition there are often valuable statistics which should be collected such as price data or information on the volumes of produce delivered to the market. As the extension officer becomes more experienced his own observations are extremely valuable. It is very important to cross check information whenever possible. Even then views will be contradictory. This is because an individuals' views are often narrow and prejudiced. The extension officer should aim to become an expert in the market's requirements and become the eyes and ears in the market on behalf of the farmers he or she represents.

Price Determination

Price information is a critical part of market research. Prices can normally be used not only to show how much the farmer should receive for his produce but, in a free market, also what worth customers put on the crop.

The problem is that prices of horticultural crops vary from day to day. Past prices are no guarantee of the prices which will be obtained in the future. Analysis of them will give guide prices for the future to be used in budgeting. Perhaps more importantly price analysis will also give an indication of the typical seasonality of prices and therefore show the best time to market crops.

There are normally two sources of price data, official price data and typical prices provided by the wholesalers. Official data is often recorded inaccurately and wholesalers must be considered an unreliable source of price data.

Nevertheless, provided the extension offficer understands the weaknesses of price analysis in predicting future prices, the work needs to be done to give a broad idea of typical prices and seasonal price patterns.

By analysing price statistics it is possible to show price patterns and predict months when high prices can be achieved. Analysis shows whether prices are going up faster or slower than inflation and, using these figures, it is possible to predict likely future prices. As explained earlier, past prices do not necessarily reflect future prices. Horticultural marketing is like gambling: careful analysis helps improve the odds in favour of the grower However statistics are very often inaccurate. Some of the common reasons for inaccuracy are that:

- the official prices that are gathered are normally those at the peak of the market sales, rarefy do they reflect the much lower prices of sales made at the end of the day;
- the staff collecting the information are rarely trained and often have no appreciation of the potential uses of the information. Sometimes there are inconsistencies between the collectors. For example one person may collect the top prices for first-class produce while another gathers average prices;
- sometimes the prices quoted are what the official price should be not the price that the produce is actually sold at.

The extension officer should be aware that the prices collected are the prices obtained by the wholesaler or commission agent not what the farmer will get. Commissions, margins, market charges and marketing costs will have to be deducted.

Price data must be viewed with scepticism. Most confidence can be placed in the general price patterns but not necessarily the actual figures. With experience, when price patterns are drawn up on graphs, it is possible to learn a lot about the supply patterns to the market by remembering that in general high prices indicate low volumes and vice versa. This understanding can be helped by drawing graphs of the supply pattern.

Price data should also be collected from middlemen, be they orchard contractors, commission agents or wholesalers. This is essential in order to compare and check official price data. In the absence of official statistics, such people are the only source of price information.

For commercial reasons middlemen will often give distorted price information. For example, a middleman who sells on a commission basis will tend to exaggerate prices as it is in his best interests to encourage sales. Conversely a middleman who buys produce will often underestimate prices to discourage further supplies if he thinks this will provide competition.

Most middlemen make their money in spot trading, that is buying and selling produce at a profit on a daily basis. Although they will know the prices over the last few days they are usually unaware of how prices have changed over the last few years. They are normally very poor at identifying long-term market trends.

Supply and Volume

The quality and detail of production statistics varies considerably from country to country. Regional or provincial statistics giving crop areas can help to identify the major producing areas and their relative importance. This is useful in identifying competing areas. Some statistics will also give tonnages of crops produced. These figures are notoriously unreliable as they have to be based on crop areas multiplied by an estimate of yield. Yield figures used are often little more than guesses.

Quality Requirements

Quality in terms of grading, packaging and presentation can significantly affect sales and prices but quality requirements vary considerably.

Quality Specifications

Occasionally export markets will have stipulated grading standards which have to be met by the exporter. In practice these are normally minimum standards and, because of the transport costs involved, only top quality produce is exported. It is very important to establish directly from the potential market what quality standards are required, how produce should be presented, what size and type of packaging is preferred and what price differences exist between the various grades? Generally this information is readily available from middlemen and wholesalers. Sometimes they will give actual size requirements and may specify particular varieties or appearance characteristics which they favour. From talking to the wholesale market the extension officer should attempt to draw up with them a crop quality specification.

Horticultural quality is very difficult to communicate merely through words. It is a tremendous advantage to be able to show photographs of what is considered to be good, average and poor quality produce.

Very often there are differences between markets in their demand for quality produce. For example, it may be possible to supply the major cities with the highest quality produce while the medium quality produce is sold to nearby towns. The poorest quality is sold in the villages and outgrades are used as animal feed.

Packaging

The principal purpose of packaging is to reduce damage in transport. Another purpose is to keep the produce in a sensibly sized unit for handling and marketing purposes. In addition, good packaging can contribute to the attractiveness of the produce and help to promote sales. Packaging is, however, expensive. Indeed in field vegetable production in Europe the single most expensive production cost is very often the packaging material. In practice any recommendations on the introduction of new packaging material must carefully weigh the additional costs against the likely benefits.

While visiting wholesale markets it can be very instructive to examine the amount of damage caused in transport. Produce at the bottom of boxes should be looked at as well as the crops that have taken the full weight of the truck's load. Very often there is a conflict between the lorry driver's objective of transporting as much produce as possible and the problems of overloading and crushing the lower tiers of produce. Well-designed packaging needs to maximise the use of space so it is normally oblong or square. To prevent tiers of boxes crushing the lowest produce the strongest points need to be the four corners of the box. The floor of the box acts mainly as a shelf. A certain amount of air movement needs to take place through the stacks of produce to prevent the build-up of heat and gasses. Sometimes, however, if produce is to be transported in very dry dusty conditions, excessive air ventilation can become a problem.

The size of unit that the produce is packed in is very often determined by how the produce is to be sold. Major products are normally packed in larger containers. Smaller volume commodities have to be packed in small units because the retailer would be reluctant to buy more produce than he thinks he can sell. Recently there have been significant developments in packaging in consumer-sized units. In the Arabian Gulf market, for example, where large individual purchases of produce are common, the more progressive exporters have now started packaging a portion of their produce in retail units, e.g. grapes in three-kg boxes and five-kg nets of onions. In other markets there have been developments in pre-packaging tomatoes, cucumbers and sweet peppers into one-kg nylon nets.

THE MARKETING SYSTEM

One should aware of the best way to work within the existing marketing system. This involves building up an understanding of how produce is distributed and sold. The price relationships between the different sales points in the marketing chain have to be studied. Knowledge must be gained about which companies in the distribution chain have reputations for honesty and integrity. Finally it is important to understand how growers can be kept regularly updated with market information in terms of prices and volumes and quality required.

Middlemen are subject to a lot of hostility, much of it unwarranted. They are generally accused of making excess profits and of dishonesty. It is sometimes not realised that middlemen perform an essential function in carrying out the marketing of produce. They are, in effect, the channel through which produce is taken out of the rural areas and money returned. Experience shows that, provided a market opportunity is identified, which is normally the responsibility of the middleman, farmers will respond by producing the crops. Only rarely is a lack of technology the critical constraint.

Provided middlemen operate in an atmosphere of strong competition it is unlikely they will make excessive profits. Clearly if this were the case numerous other businesses would be attracted and the competition would force down profits. As in all businesses there are some middlemen who are dishonest. The challenge for the extension officer is to identify reputable middlemen and to ensure that the marketing system minimises opportunities for dishonesty.

Product Distribution System

There are very many different systems of marketing fruit and vegetables. They differ from country to country and indeed sometimes from crop to crop and from farmer to farmer. The extension officer will need to fully understand the marketing system if he is going to be able to make it work for the benefit of farmers.

Some likely business people who can form the links in the production/marketing chain are:

- contractors who buy crops in the field and undertake the harvesting;
- agents, collectors, hawkers or country wholesalers who buy the harvested crop at the farm;
- wholesalers who buy at rural assembly or village markets;
- commission agents or auctioneers who auction produce in a wholesale market on a commission basis;

- wholesalers who sell produce on a consignment or commission basis in the wholesale market;
- wholesalers who buy produce from farmers at firm prices and sell at the wholesale market for their own account;
- exporters and importers;
- secondary wholesalers who buy at the wholesale market and transport the produce either to sell to retailers or at another wholesale market where prices are higher;
- semi-wholesalers who are located near the wholesale market and sell produce by the box either to small retail businesses or directly to consumers.
- retailers who sell to the final consumer such as street hawkers, stall holders, retailers, greengrocers, supermarkets;
- catering establishments, food processors.

Around 70 percent of the produce is sold as a standing crop to an orchard contractor. Normally a number of contractors will bid for an orchard. One third of the price is paid three months in advance and the other two payments are made during harvesting and at the end of the crop. The orchard contractor undertakes the harvesting, grading, packing and distribution of the crop and may hold the produce in cold storage. This system has a number of advantages:

- the farmer is partly paid in advance;
- the farmer need not worry about organising labour for harvesting;
- the contractor's labourers are skilled in harvesting, grading and making up the wooden boxes on site;
- the contractor will normally undertake the harvesting on a number of small farms which will achieve economies of scale in terms of transport;
- the contractor is a specialist in marketing and understands the requirements of the commission agents and wholesalers.

Marketing Margins

The proportion of the final retail price that is resumed to the grower arouses much emotion and discussion. Calculating the margins in the marketing chain can be a difficult exercise because they vary depending on the retail price of the product, its perishability and the marketing costs, particularly those for transport and packaging.

It is often mistakenly assumed that the difference between the retail price and the wholesale price is the retailer's profit. This is wrong because it fails to take into account the fact that produce is often sold at different prices and that some produce is downgraded or even wasted entirely.

It is also a mistake to assume that the margin or mark-up, which in this case is nearly 4 LE per box or 0.22 LE per kg, is purely profit. The margin has to cover transporting the tomatoes and any salaries, rents, taxes and investments in equipment as well as to allow the stall owner an income and an opportunity to put money aside to cover periods when losses are incurred.

It is of particular interest to work back from the wholesale price to the farm gate pace. To do this the extension officer will need to establish what margins are calculated in the wholesale markets. Information like this is highly sensitive and will need to be rechecked from different sources, including farmers, contractors, commission agents, wholesalers and retailers.

To do this properly the extension officer will need to retrace each step in the marketing chain and establish from each middleman his buying and selling pace. Typically, margins are greatest when the middleman pays a firm pace and actually takes ownership of the produce. This is because he is taking the Ask and Ask and profit are closely linked. When produce is auctioned or sold on a consignment basis, the middleman, be he a wholesaler, importer or commission agent will hold back a percentage commission, normally from the seller but sometimes also from the buyer.

Using the example of a tomato crop in Egypt again, the method of calculating the return to the grower is set out. As has been shown, accurate calculation of margins is extremely difficult. All too often it is assumed that from a simple comparison of an individual retail price, wholesale pace and farmer pace it is possible to accuse middlemen of excessive profits.

Role of Wholesalers and middlemen

An important part of the extension officer's work is to identify suitable and reputable middlemen as trading partners. This involves finding out which companies are best equipped and most prepared to trade in the produce from his area. Secondly, he should find out whether these companies have a reputation for integrity and honesty. Discovering this information involves not only having meetings with possible trading partners but also, in effect, taking references about their reputation from other traders.

There is often a shortage of simple company information on potential trading partners and yet it is crucial when growers are planning to supply a new market or start marketing. The top priority should be honesty, but it is also important to identify businesses which are appropriate to the type of commodities which are planned to market and the scale of production which you envisaged.

Information Services

It is absolutely vital that market news information is accurate and rapid. In practice the most important method of communication is the telephone but few farmers in developing countries have access to phones. The telephone can be one of the best investments for a horticultural producer; the additional returns may outweigh his costs many times. In the absence of a telephone link a well-run radio market news service broadcasting price and state-of-supply information has a number of advantages. As almost all growers have transistor radios the information is available to all. Growers can respond to market opportunities by diverting produce from one market to another. Ultimately, price differences between markets are reduced, which provides benefits to the consumer in terms of price stability and better continuity of supply.

Normally wholesalers and middlemen have an advantage in negotiation by being better informed on market prices. The growers' negotiating position is improved by having ready access to price information. This is particularly important when there is little competition between buyers, e.g. when an individual contractor bids for a fruit orchard on a farm. The extension officer needs to consider ways in which price information can be made more available. Some techniques used include publishing information in newspapers and chalking up typical prices in markets and village meeting places.

Extension officers should also ensure that the market information provided is relevant to the farmers' needs. Prices in distant markets which cannot be easily served by farmers are of less relevance than prices in nearby villages and town markets. These "local" prices are often not broadcast by radio stations. The extension officer should collect and disseminate these prices back to the farmers. This can be done by word of mouth or by placing notices on bulletin boards at locations regularly visited by farmers e.g. market places, chief's offices, input supply points, etc.

One danger which the extension officer should be aware of is growers not understanding the reason for the difference in price between the farmer, the wholesaler and the retailer. It needs to be explained that distribution and marketing involves costs and business risks and prices have to cover the salaries of those involved in the distribution chain. Unless these costs are covered the marketing system will collapse.

DECISION MAKING

Credibility can be a serious problem for extension officers. Farmers will often be skeptical of advice from someone they consider to be inexperienced in practical matters. Growers may be suspicious of the motives for providing free advice. They are understandably reluctant to accept untried advice, particularly when they will suffer financially if it proves to be wrong. Furthermore, it is often those who most need good advice who are the most difficult to contact. These are the small, poor farmers without transport who are usually the most conservative. Wealthy farmers will probably have the transport to visit extension officers and are often more appreciative of extension advice. They are, however, those that least need assistance.

The challenge to the extension officer with special responsibility for marketing is firstly to decide how the marketing problems of the area can be solved. Secondly, he or she needs to think through the best way to get advice or plans across to the maximum number of target farmers. Finally, the agreement and the commitment of those that will be involved in any coordinated production programme in the area must be obtained.

Identification of Problems

The ways in which problems can be solved and opportunities exploited will change from area to area. To help the extension officer take a clear overview of the area, he or she will need to identify what stage in horticultural development the region has reached. Normally the aim will be to try to introduce the next steps in horticultural progress. The best solutions to marketing problems are normally relatively simple and ought not to require any major changes in production or new technologies. Complex plans or highly innovative plans are much more likely to fail.

Three examples are given:

- If the area produces but has not before sold horticultural crops then the extension officer should be looking for ways to establish a local market. This might involve, for instance, coordinating growers to assemble their produce on one particular day of the week at a convenient location. Middlemen would be invited to attend the market. The buyers would compete with one another to buy produce so that fair prices should be achieved. As produce is assembled in volume, cheaper bulk transport to the major markets is made possible.
- If produce is only being sold into a local market there may be opportunities to start supplying more distant, major markets. The extension officer's research should have indicated what the produce requirements, prices and costs are likely to be and identified potential trading partners. He may then want to persuade growers to attempt a test marketing programme to new markets, initially of existing produce and then for new types of produce. If this proves successful then commercially sized shipments could be made. Another option would be for a local representative to take on the role of transporting and selling produce.
- In an area which is already a major force in horticultural production and marketing the extension officer may concentrate on improving the existing system.. When individual growers already have good links with the market it will be difficult to

form producer groups or cooperatives-unless there is some major problem or need. Extension advice is likely to cover generalized advice to groups of growers and specific advice for individual farmers. Potential improvements are likely to be identified at critical points along the production/marketing chain. These may involve anything from new crops and improved production practices to post-harvest techniques, better designed packaging, improved transport methods, better access to credit and production inputs and the establishment of improved market information services.

Finally, it is important to remember that there is always the danger of trying to make changes when they are not necessary. All systems are imperfect inevitably farmers will always complain that they receive too little money while consumers complain of too high prices. Proposals for change do need to be carefully thought through and any additional costs or disadvantages balanced against the advantages.

Extension Techniques

The two chief functions of an extension officer are:

- to reduce the learning time for an individual farmer to accept a new idea or technique;
- to increase the number of farmers who understand the new ideas.

In some countries certain extension officers are given the responsibility of becoming subject matter specialists in marketing. It is then their task to train other extension officers and provide specialist marketing advice. By working through other extension officers their effectiveness is increased. Generally the marketing extension officer is based in the production area. He or she must make regular visits to the markets in order to maintain contacts and keep in touch with changes in price and demand. Sometimes, however, he is based away from the production areas. In this case he will have to return regularly to the production areas to maintain contact with growers' problems and to provide relevant marketing advice.

Experience has shown that one of the most effective ways of working is to work with groups of farmers.

An extension technique, very much under-exploited, is helping farmers indirectly by providing guidance and advice to private-sector companies. For example, companies who either supply inputs or, more importantly, assist in the marketing and distribution of produce can often use timely and reasoned advice to put into motion a process which brings benefits to large numbers of growers. For example:

Table 1. Extension Techniques

- *Farmer teaches farmers.* A successful farmer explains to a group of farmers his production and marketing practices. The meeting is most effective on the farmer's own farm.
- *Demonstrations.* Practical demonstrations of techniques such as harvesting, cleaning, grading and packing, preferably taking place on a farm. Prepared samples which demonstrate the differences overtime of different handling practices can be effective, as are samples of competing produce and photographs.
- *Talks and seminars.* Possible topics include: market possibilities successful case studies, postharvest techniques, prigs assessment, market-oriented production techniques. Buyers and middlemen should be involved to talk.
- *Problem-solving techniques.* The farmer group is encouraged to identity its own major problems. The problem solving can be tackled systematicaly, by calling In specialists individually to advise the group or by forming a panel to answer farmers' questions. Alternatively, the group might be encouraged to decide their own solutions which they then implement themselves collectively.
- *Study tours.* Farmers are taken on a study tour to make their own contacts and to see the market for themselves, visit processing centres and observe how their produce withstands transportation. Farmers visit farmers in another area to exchange experiences and see new techniques. This experience alone can transform a grower's views on production and marketing.
- *Written information.* Fact sheets are prepared and distributed. These can identity potential trading partners or provide technical information on production and post-harvest techniques.
- *Market news services.* Establishing a market news service which provides regular, reliable, relevant and timely information. This may be in the form of a news sheet or a radio bulletin.

- if shortages of the required input supplies are restricting market-oriented production, shops selling agricultural equipment may be persuaded to buy in the necessary materials and even advise growers on their correct use;
- if middlemen or traders can be persuaded that there is a business potential in marketing produce, they can be very effective in establishing workable marketing systems. They can supply seeds and packaging material, give guidance on produce presentation and provide credit for farmers;
- advice and encouragement could be provided to a packaging manufacturer to improve design or use better materials;
- a local transport firm could be assisted to establish a produce collection service.

Leading local farmers strongly influence the decisions of other farmers, e.g. to coordinate their production and marketing. They will generally need to be persuaded of the potential advantages to themselves of improved marketing practices. The extension officer should beware that unless the wealthy farmer is genuinely concerned for the community's welfare he may try to suppress the planned changes. He may do this because he does not want other farmers to become better off. Alternatively, he may see an advantage in carrying out the scheme on his own.

The extension officer may also find it valuable to work with public-sector organizations such as:

- agricultural banks, to improve their credit service to growers;
- farmer cooperatives wishing to undertake input supply and marketing activities;
- research institutes, to ensure they tackle farmers' market-oriented production and post-harvesting problems.

The scope of the work clearly extends far beyond advising farmers. This is because marketing is only as successful as the weakest link in the production/ marketing chain.

Agreeing on an Action Plan

In this section we have considered four potential activities of the extension officer. The first involves giving advice to an individual farmer. This should be a low priority as it is an inefficient use of time and larger-scale farmers obtain the most benefits. cThe second involves providing marketing advice to farmer groups, particularly through the methods of mass extension. This has the advantage of reaching a large number of growers and allows coordination of farmers' activities and cooperation in marketing. The third technique involves providing advice or information to critical individuals, organizations or private-sector companies in the marketing chain whose actions can have a beneficial effect on marketing. Finally, and perhaps most ambitiously, an extension officer may decide it is necessary to attempt a project approach to developing horticultural marketing. Marketing is normally achieved by a series of interlinking stages and coordination between the stages is essential. A project approach is one which involves coordinating the activities of a number of different intermediaries in a marketing chain. It may involve a group of farmers assembling their produce at one point so that it can be transported in bulk to the market.

More complex schemes could involve ensuring a supply of inputs, providing growers with production advice and negotiating contract terms with a buyer, be he a food processor or exporter.

Extension officers can have a credibility problem. One way of overcoming this is to achieve a good reputation by successfully resolving some smaller problems. A second way involves securing influential support for the scheme, particularly from farmer leaders or marketing companies. Sometimes them is a reluctance for people to implement someone else's ideas enthusiastically. A clever advisor sometimes overcomes this problem by not revealing his project plans to the individual he considers to be the most important in the scheme. Instead he provides him with the information on the problem and then in discussion leads the individual to come up with the same (or a similar) solution. The individual then thinks it is his own idea and has the enthusiasm and commitment to ensure the plan's implementation.

The extension officer's role is then to support the individual, to coordinate the activities of the different parties involved and to chase up the progress of the project.A planned project approach to horticultural marketing development will increase the chances of genuine improvements being made. It is important for the extension officer to have a clear mental image of the desired outcome and successfully communicate that objective. The project must be understood by all parties, if they are going to be able to work effectively together.

Business management experience has shown that targets, such as tonnages to be shipped or selling prices, are important too, as they provide challenges to the parties involved and can be used to monitor the progress of the project. This technique is called management by objectives. However, no matter how good the preliminary work has been, when a plan is put into action the unexpected will happen. Allowances for the unexpected should be made. It is advisable to start the project with a pilot stage so that mistakes can be made on a small scale and reamed from. Furthermore the project must be flexible so that changes can be made in the light of these lessons. Inevitably the project's critics will try to emphasize any problems; most successful projects will have had to face problems, particularly in the early phases, and their success is often a measure of their ability to learn from and overcome difficulties.

IMPLEMENTATION OF ACTION PLANS

At each stage in the production/marketing chain there are possibilities for improvement. The extension officer will have to identify the main problems and concentrate his efforts on those areas.

Pre-production Advice

Input supply

Both quantity and quality of crops produced are affected by difficulties in obtaining inputs. The correct planting material is particularly important. Often consumers have strong preferences for particular varieties of vegetable. In much of the Middle and Near East, for example, there is a preference for plum-shaped tomatoes. Red-coloured fruits, such as red sweet apples, are sometimes much preferred to green and golden varieties. Growers'marketing margins can be improved by ensuring the supply of the correct planting material. The extension officer's role is to advise the nurserymen and seed suppliers on which varieties they should supply, and the farmers on which varieties to plant. Pest and disease damage will seriously reduce a crop's price and its potential shelf life. Sometimes these problems can be solved by the correct crop protection practices. An example of this is the introduction of spray programmes to control scab disease on apples in Kashmir, India. A crucial step in the successful introduction of this programme was ensuring that agricultural chemical shops had the recommended materials available and could advise growers on their use. The extension officer should be alert for opportunities where farmers themselves can become input suppliers. Farmers can also provide contract services, e.g. for soil cultivation or crop spraying.

Finance and credit

A critical production constraint is often shortage of funds. Broadly speaking the potential sources of funds can be divided into two-formal and informal. Formal sources are mainly banks. In general, the interest terms are reasonable, but requirements for security and slow bureaucracy often limit loan effectiveness. The extension officer can help with the supply of production credit by providing the bank with cost-of-production data and likely returns.

Loan agencies have different rules as to what proportion of the production costs can be advanced. Some will only cover the cost of inputs, while others will include some or all of the labour costs. It is an advantage if the marketing cost of transport and packaging can also be covered. It is interesting to note that in this example harvesting and marketing costs amount to 67 percent of all costs and that packaging is the single largest item.

Formal sources of credit are also extremely valuable to cover long-term investments, such as planting and establishing fruit orchards and investment in production equipment.

Some banks offer marketing loans where money is provided to cover the harvesting, transport, packaging and even storage of crops. The amount advanced can be calculated on the basis of the harvesting and marketing costs, as set out above. Sometimes they cover a portion of the estimated wholesale value of production. The value of these loans is that they can free the grower from the marketing restrictions imposed when borrowing from a middleman. These normally commit the grower to selling all his produce via the one middleman.

Where growers are under contract to supply produce to an agribusiness, e.g. food processor or exporter, production loans based on the hypothetical value of the crop can be introduced. For example, if a farmer is growing one acre of tomatoes for a food processor and he would normally harvest five tonnes per acre with a contract price of

$0. 25 per kg, his income is expected to be $1 250. Under a loan hypothecation scheme the bank or processor advances a proportion of the expected income as production credit, say 25 percent or in this case $312.5, without seeking any additional security. Loan recovery can be made from the agribusiness from the money it is returning to the grower.

Although informal credit sources include family members, friends and input suppliers, the most important source of credit is often the middleman or commission agent.

The role of a wholesaler as a banker is much misunderstood. Common criticisms are that very high interest rates are charged and that growers who have borrowed money are forced into selling their produce at low prices. In some cases this is true but it is not as common as is assumed.

Generally the produce is auctioned in an open market so the grower should receive a fair price. The loans are often made on the basis of close kinship links and are free of time-consuming bureaucracy. Loans are recovered simply by deducting the money advanced from the sales.

The main disadvantage of this system is that the grower has no marketing flexibility. In the case of poor prices he cannot switch to another commission agent or market.

In the absence of an effective and suitable formal credit agency, marketing middlemen can be a useful credit source, particularly if finance is a major constraint to starting production.

Production Planning

Extension officers should be able to advise farmers on planning their crops. Although important criteria such as labour availability and crop rotations will have to be taken into account the key approach will need to be market oriented production. This means growing crops for which there is likely to be a demand and which will probably be profitable.

Individual crops selection

The extension officer should calculate potential or net returns of the major alternative horticultural crops in the region. This will establish which crops are likely to be the most profitable. The market research that he or she will have undertaken should have shown whether any of the local crops have a comparative advantage. The produce may have advantages in terms of price, quality or seasonality over competing crops from other areas. Research should also have shown what varieties are favoured and the best time to supply the market. The extension officer will need to translate the market requirements into practical recommendations for farmers, covering areas such as:

- best varieties;
- sowing dates, e.g. whether to extend the period of supply or aim for a particularly high priced period; whether to avoid times of oversupply;
- other techniques to extend production into high priced periods, such as late or early varieties, transplanting techniques, polythene tunnels, irrigation;
- techniques to improve quality,. such as optimum fertilization, crop protection, pruning, irrigation, weather protection.

New crops or the introduction of new technologies or production techniques should always be undertaken, initially on a small-scale trial basis.

Selection of range of crops. In practice crops to be planted are chosen for a combination of reasons. The most important is that there is likely to be a good market demand and the crop can be grown profitably, as outlined in Table 12.

Generally it is advisable to grow a range of crops as this lessens the impact of an eventual crop or market price failure. If possible some crops could be grown under contract. Very often individual growers will have preferences for crops which they feel comfortable growing and/or which grow well on their land.

As agriculture develops farms typically become more specialized, concentrating on fewer crops. Growers become more skilled, although experience has shown that growers can rarely be experts in more than three or four crops.

As explained earlier, the most potentially profitable crops are often the most risky. It is useful to have a cropping system in which risky crops are balanced against more reliable income-earning crops.

Small farms generally have more labour available per acre. They can take advantage of this by concentrating on growing labour-intensive crops.

These are crops which cannot be harvested mechanically and may also require transplanting, pruning, hoeing and multiple-hand harvesting.

Investment Advice

Business and investment advice can also be part of an extension officer's work. Often farmers are tempted to make investments which are expensive and do not improve the viability of their farm significantly. The priorities for investment decisions should be:

- investments that can assure a farmer of an income;
- investments that can increase incomes by improving prices;

- investments that can improve incomes by increasing yields;
- investments which reduce production costs;
- investments which can create additional income sources.

Investments which can improve yield stability are:

- those which protect against pests and diseases, such as sprays and a sprayer,
- those in irrigation and technology which reduce the adverse effects of the weather, e.g. through protection of nursery plants.

Investments like these are particularly important in horticulture because in seasons when yields are low as a result of poor weather or pests, prices rise significantly. It is the growers whose yields arc least affected who make the most profit.

Investments which improve prices can be:

- those in technology which facilitate off-season production, such as polythene tunnels and improved transplant technology;
- those made directly in improving marketing such as grading facilities, on-farm storage, pick-up trucks to transport produce and even telephones to improve market communication.

Post-production Advice

The timing, technique and conditions at harvesting can significantly affect prices.With some crops, harvesting can be undertaken early to take advantage of high-priced opportunities, e.g. cabbage harvested as spring greens, young carrots sold in bunches, green plums and new potatoes. Exploiting these short-term market opportunities requires a close link with the market.

Shelf life and long-term storage is affected by the maturity of the crop at harvest. The storage characteristics of root vegetables are generally improved by only harvesting fully mature crops. Examples are sweet potatoes, tannia, carrots, onions, garlic, potatoes and yams.

Harvesting of cassava tubers can start between seven to ten months after the planting of cuttings, depending on variety. Cassava does not store well and on small family farms the largest tubers are harvested first, without cutting the stems. The small tubers are allowed to grow on. The production of cassava roots is at its highest 18 to 20 months after planting.

Long-term storage of cabbages depends on the cultivar and growing conditions. Cabbages suitable for storage are normally slow growing, large types grown under cool conditions and harvested at the correct stage of maturity. Harvesting should take place after the head has formed tightly and before the outer leaves start to die or the head shows any sign of splitting. At the correct stage of maturity freshly harvested cabbage heads should squeak when rubbed together.

Melons are another example of a crop where the timing of harvest is crucial for quality-too early and the full sugar content is not developed, too late and they lose sugar and become soft. Cantaloupe melons should be harvested when the fruit separates easily from the plant. If only part of the stem pulls off then the fruit is not ripe and will never ripen to a full flavour. Honeydew melons and watermelons do not separate from the stem when mature. A honeydew melon is ready for harvest when the fruit is well filled out, there is just a hint of green and the surface is covered with fine hairs. Watermelons should be harvested when the ground spot is pale yellow and the fruit gives a hollow sound when hit with the knuckle.

Some fruit have to be harvested when they are not completely ripe in order to transport them to distant markets This is particularly true of fruits which are not suitable for long-term storage but need their shelf life maximized. Examples are bananas, pineapples, mangoes and avocados.

In the case of bananas the fruit is generally harvested when still green and only at between three-quarters full (80 days from shooting) to high three quarters (90 days from shooting). The longer the period of transport the thinner the fingers at the time of cutting. Maximum storage of two to three weeks is achieved by keeping the fruit at 12-14°Cin 85 to 95 percent relative humidity. On arrival at the market bananas are ripened either in special ripening rooms or by allowing the ambient heat to trigger the release of the ripening gas, ethylene. For local marketing the fruit is harvested when fully mature but before ripening has started.

Avocados and mangoes will generally ripen during transport. Pineapples for local consumption or canning are normally harvested when the fruit has yellowed up by 25 to 50 percent. For distant markets harvesting should take place when the first hint of colour change has been observed at the basal end.

The optimum harvesting stage for most crops will depend not only on the climate and distance to the market but also on variety and growing conditions. In individual cases, when new distant markets are being explored, experiments should be carried out to find the best stage to harvest fruits, by sending samples at different degrees of ripeness and assessing which is most favoured.

For fruits which are suitable for long-term storage, such as apples, pears, citrus and grapes, there are significant differences between cultivars, growing region and sometimes season in the optimum harvesting time. For example, maturity indices of citrus are based on juice content by volume, total soluble solids in the juices and the solids: acid ratio, according to variety and market. In general, fruits should contain at least 40 percent by volume of juice and the total soluble solids should be in excess of eight percent. Specialist skills and techniques are necessary which are beyond the scope of this book. Apples for long-term storage should be picked when fully mature but not fully ripe. The extension officer should call in the necessary expert assistance if long-term crop storage could significantly improve farmer incomes.

Harvesting and Quality

What is often not understood by growers is the effect of their harvesting and handling on the quality of the produce in the market. Once a fruit is plucked from a plant or a root or leaf vegetable is harvested, it is cut off from its source of food and, particularly, water. The effects of poor treatment normally show themselves some days later, when the produce is being presented for sale or is in storage. Poor treatment has two effects; firstly the price is reduced and secondly, in the long term, the reputation of the production area is diminished (again tending to result in lower prices).

An improved system of harvesting and handling produce will result in a product with better appearance and shelf life. In general, prices for the produce will be improved but sometimes the system has to be changed to ensure that the price rises are passed back to the grower.

Ideally harvesting should take place when the crop and the climate is coolest and the plant is most turgid, i.e. has the highest moisture content. This is in the early morning. In practice other criteria also have to be taken into account. For example, the dew should be dry on citrus end the latex flow of mangoes is at a minimum a/mid-morning. Harvesting also has to take into account labour availability and when collection will take place, to minimize the time produce is left standing in the field.

Harvesting Techniques

On high trees fruit can be harvested with a hook and a catching bag on a pole or similar harvesting aid. This prevents fruit falling to the ground.

For other crops knives and clippers can improve harvesting practices because they can cut through fibrous tissue, stems and leaves can be trimmed and clean cuts reduce the likelihood of infection. Tools like this are used for harvesting lettuce, cabbage, sweet pepper, egg-plant, honeydew melons and banana. In the case of bananas a slight cut is made in the upper pseudostem to allow the bunch to ease down gently. Individual hands are cut off from the bottom hand upwards.

Leafy vegetables are harvested by cutting the plant with a sharp knife as close to the root as possible. Uprooting results in soil coming into contact with the produce.

Bulb crops such as garlic and onions are harvested by pulling the leaves at the neck and then cutting the leaves about 3 cm from the bulb.

Occasionally diseases can be transmitted from plant to plant. Tools should be cleaned often and, when virus diseases are a problem, knives should only be used for trimming not for cutting the fruit from the plant.

Many fruits are harvested by hand, e.g. apples, citrus, papaya, peppers, tomatoes. The fruit should beheld by the palm of the hand not by the fingers. Whenever possible the harvesting should be carried out by plucking the stem, e.g. with strawberries, fine beans, peas.

Tuber and root crops are normally harvested with forks or hoes. The digging should start some 15 cm (6 inches) away from the tease of the plant. In general, it is preferable to lever and pull the roots rather than attempt to dig the roots out. Harvesting is easiest when the soil is relatively dry as both damage and the need for washing is reduced.

Field Containers

Picking bags or baskets attached to the waist of the picker enable both hands to remain free. The crop damage associated with moving sacks of produce through the field is reduced. With picking bags it is preferable to be able to release the bottom so that the produce can be let out gently, rather than upending the bag.

Baskets or boxes with sharp or rough edges should either be avoided or lined with paper or leaves. Damage is often caused by transferring produce from one container to another. If possible, produce should be harvested into the container in which it will be stored or transported.

Harvesting System

With highly perishable produce damp cloths can be used to give protection against the sun's heat. Field containers should be removed to a shaded area as soon as possible. Some leafy vegetables may be sprinkled with water at intervals to maintain leaf turgidity. Field assembly points, such as a shadehouse made out of natural materials or a canvas tent, should be used in order to keep the produce cool and allow ventilation.

In general, the quality of fruit and vegetables cannot be improved after harvest. However, the more careful the handling the slower is the deterioration in quality. Containers must be emptied carefully to minimize drop heights and fruit-to-fruit damage. Containers should be periodically cleaned.

Curing and Drying

Bulb crops such as onions and garlic can be dried in the field by being spread one layer thick over about six dry days. Alternatively drying can take place in stacked shallow trays under cover. The aim is to harden the outer scales and remove moisture from the neck in order to extend storage and marketing life.

Most root crops (but not cassava) respond to warm moist conditions after harvest by thickening and hardening their skins. This provides protection against dehydration and infection. Wound healing occurs. This is called curing and it significantly improves storage life. Curing can be carried out in tropical areas at little cost by stacking the produce in conditions where temperature and humidity are allowed to rise to 25 to 35°C with a relative humidity of 85 to I 00 percent for one to seven days, depending on crop and variety.

Trimming and Sorting

Cabbages, cauliflower, chinese cabbage and lettuce will have their outer leaves trimmed, except for three or four wrapper leaves, to give some protection to the head. Long stalks attached to the fruit, as in citrus, should be cut as close to the fruit as possible to prevent damage to other fruit.

Provided the market wants graded produce and is prepared to pay for it then selection and grading are justified. The additional prices must cover the additional costs. Buyers may specify grading standards, particularly in the export market where international standards may be enforced. Produce for longterm storage should be disease and blemish free and therefore needs to be sorted. When transport is expensive it is often only justifiable to send top quality crops. Produce is generally separated according to quality criteria, it may also be graded according to ripeness or colour and size. The crop is then normally packed into different containers. This facilitates marketing into different markets.

Grading and packing is often carried out on the ground under the shade of a tree. This is both unhygienic and inefficient. Specialist grading areas or sheds are generally open-sided, with tin or preferably thatched roofs. Grading while standing or sitting at tables enables people to work faster Tables covered with polythene sheeting are easy to clean and the sheeting can be replaced cheaply. Lighting should be good. Tin roofs can be painted white to reflect heat while water trickled down the outside of a shed helps reduce the heat inside the building.

Packaging and Presentation

The two main functions of packaging are to help prevent mechanical damage and to sort the produce into an acceptable size for the market and for handling. Good packaging can also enhance the attractiveness of the produce.

The four main types of mechanical damage are cuts, compressions, impacts and vibration rubbing.

Care in harvesting and handling will help eliminate cuts and wounds. Lining of packaging material with paper or leaves can also prevent damage.

Compression bruises can be restricted by using containers which are strong enough to withstand multiple stacking. The packaging materials need to be particularly strong at the vertical corners. The packaging should also be shallow enough to prevent the bottom layers of produce being damaged by the weight of produce above. Cartons must not be overfilled. Damage is caused by the full weight of the pile of produce pushing down on the top layer of fruit or vegetables, causing the weight to be transmitted to the lower layers. Impact damage and bruising can be the result of shocks in transport or dropping. This may occur either because each package is small enough to be thrown or too big to be easily handled.

Vibration damage generally occurs during transport; vibration being transmitted through the produce. This kind of damage can be significantly reduced by achieving a balance between preventing the produce from moving within the packaging and forcing the produce together. Fruits are prevented from rubbing against one another by the use of cellular trays, individual wraps or cushioning pads. An example is paper and straw used to separate layers of apples. Alternatively, the box is gently shaken to settle the produce and then the space created is filled.

Reference was made earlier to the size criteria of packaging. The largest size should not exceed 50 kg as this is the maximum weight which can be easily handled. Below that the size specification will depend on the customers' requirements-be they the retailers or consumers.

When attempting to introduce new types of packaging, the extension officer's first task is to compare its cost with the existing system of packaging. Subsequently he should monitor trial shipments and then carry out a cost/benefit analysis using actual, rather than theoretical, figures. The key is to select costeffective packaging which is appropriate to the demands of the market. Produce packaging materials can be conveniently divided into six classes.

Locally available natural materials, for example baskets woven from bamboo, willow or cartons made from thin strips of wood or rushes. Typical problems associated with these materials are:

- poor rigidity and design, which prevents multiple staking;

- sharp edges, which can cause bruising or pierce produce;
- inefficient usage of transport space, which increases costs.

There are, however, a number of advantages with using locally available materials. Material costs are low. Both jobs and incomes are created for local businesses who make the packaging. Local sources of packaging also make it easier to ensure its timely arrival.

The benefits are such that in the first instance the extension officer should try to develop and improve on the use of existing local packaging. This may be achieved by new designs or through improvements in the handling system.

Wooden boxes and trays are widely used throughout both the developing and developed world. They are strong, rigid and can be manufactured locally as well as recycled. They can also withstand refrigeration. However:

- wood is often not available or is very expensive;
- boxes are not designed properly, resulting in poor stacking and ventilation characteristics; or
- in an effort to save wood the boxes are too deep, resulting in damage to the bottom layers of produce.

Improved design is particularly likely to result in both savings in wood and reduced crop damage. The European produce tray has been successfully introduced in a number of countries. Critical design features of this tray include:

- standard box sizes, particularly length and width, to facilitate stacking;
- using thin strips of wood for the floor and part of the sides but especially strong wood at the vertical corners, as these have to support the weight of the stack;
- a gap between the sides of the tray and the floor of the next tray, allowing for ventilation;
- no lid but paper placed on top of the produce to reduce the effects of dust, evaporation and to minimize pilfering.

Shallow trays are used for easily bruised crops such as tomatoes, peaches, grapes and mangoes. Deeper boxes are used for apples and citrus. Larger but flimsier boxes are often used for cabbages and cauliflowers. In these developments the extension officer needs to work closely with the local box manufacturers. Again, pricing of packaging per kg of produce needs to be compared with that of existing packaging and should be tested before commercial introduction.

Fibre board or corrugated cardboard are increasingly being used, particularly in developed countries. There are a number of very cleverly designed boxes which can be copied. The boxes are light and easily printed and can be made to look very attractive. However:

- they are expensive and cannot be recycled;
- they need to be waxed if they are to withstand long-term cold storage;
- they need to be manufactured by large factories creating employment in the cities, often at the expense of rural jobs;
- the raw materials often have to be imported.

Recent design improvements include boxes that are made from a combination of wood, for structural strength, and cardboard. Plastic has also been incorporated in designs, particularly to increase strength at the corners.

Plastic containers are expensive and generally have to be imported. They are so expensive that they have to be recycled and are mainly used as field boxes or to supply a regular outlet such as a factory or supermarket. Some polystyrene packaging is now being used for non-recyclable containers.

Bags and nets are cheap but provide no protection from damage. They can be used to package suitable produce like onions and potatoes into convenient units for handling and marketing. Plastic and paper is often used as lining or wrapping for produce.

Printing, packaging presentation and brand names can all add value to produce but only in markets where consumers are wealthy and appreciate aesthetics and image. For example, in the produce markets of the Arabian Gulf multi-coloured printing is common because it has been observed to increase returns. In contrast most of the African and Asian markets are insensitive to the quality of packaging material used, if indeed packaging is used at all.

Much emphasis has been given to improved handling, grading and packaging. Improved prices may be expected because the market will respond to the quality of the individual consignment. In the longer term, premium prices can be obtained by establishing an identity and a reputation as a consistent high quality supplier. An extension officer may help to achieve this by:

- drawing up minimum grading specifications for a top quality grade;
- training farmers in grading to this standard;
- agreeing on a name, symbol or 'brand' image which can be printed on to the packaging;

- asking farmers to sign an agreement that they will only send the correct quality produce under the brand name;
- getting each grower to mark branded packages with his own identification mark so that any produce which is below specification can be traced back to the grower.

This top quality produce would be sold into high-priced, quality-conscious markets. This is called creating a brand image. There are many examples of areas which always obtain premium prices because of their reputation for supplying good quality produce. It should be remembered that it can take years to establish a good reputation but it can be quickly destroyed by one or two poor consignments.

Storage

Produce can be stored for both short-term and long-term purposes. Short-term storage is used to provide flexibility in marketing, for example when awaiting transport or because buyers are not immediately available. Most horticultural crops are perishable end can only be stored for a few days. Only rarely is it worthwhile storing crops of this nature to await higher prices. Storage will reduce quality and shelf life. It is costly and, in most instances, when the produce is withdrawn from storage it has to compete in the market against freshly arrived produce,

A few crops are adapted for long-term storage. These can be held in stores well beyond the normal harvesting period. In turn, higher prices can normally be obtained and greater volumes of produce sold. Often in the case of cold storage the successful stores are located in urban areas because:

- produce can be released immediately on to the market in response to high prices;
- facilities can also be used for other crops, e.g. apples in the winter, citrus in the summer and other produce such as butter.

Refrigerated storage is much emphasized in the literature but extended shelf life can be achieved without investment in expensive equipment. In practice the quality of the produce and humid, shady conditions are higher priorities.

Ventilated stores in the right conditions with good management can be extremely cost-effective. Ideally they require cool night temperatures. The building should be protected from the sun's heat by such techniques as shady trees, painting the building white and double-skinned walls. The building should be positioned to intercept the prevailing night time winds. When the ambient air temperature falls below that of the produce, normally at night, the air is allowed to flow through the stored produce by the opening of louvres. This process can be automated and fans can be used to increase air flow rates.

Table 1. Storage life and recommended storage conditions of crops suitable for long term storage

Crop	Degrees (Celsius)	Relative humidity (%)	Storage life (Months)
Fruits			
Apples ** +	0 to 4	90 to 95	2 to 6
Fresh date	0	85	1 to 2
Coconut	0 to 1	80 to 85	1 to 2
Grape ** +	-1 to 0	90 to 95	1 to 4
Persimmon	-1	90	3 to 4
Kiwifruit +	-05	90 to 95	2 to 35
Orange ** +	0 to 4	85 to 95	3 to 4
Pear ** +	0	90	2 to 5
Lemon (coloured) +	0 to 4 5	85 to 90	2 to 6
Mandarin	4 to 6	85 to 90	1 to 1 5
Mangosteen	4 to 5	85 to 90	1.5 to 2
Lemon (green)	10 to 14	85 to 90	1 to 4
Casaba melon	9 to 13	85 to 90	1 to 1 5
Honeydew melon	9 to 13	85 to 90	1
Vegetables			
Cabbage **	0	95	1 to 3
Carrot (topped) ** +	0	95	5 to 6
Celery	0	95	1 to 3
Garlic +	0	65 to 70	6 to 7
Leek	0	95	1 to 3
Onion (dry) ** +	0	65 to 70	6 to 8
Parsnip	0	90 to 95	2 to 6
Turnip	0	90 to 95	4 to 5
Potato +	4 to 6	90 to 95	4 to 8
Ginger +	13	65	6
Pumpkin	10	to 13 50 to 75	2 to 5
Sweet potatoes +	13	to 16 85 to 90	4 to 7
Yam +		16 85 to 90	3 to 5

** Dependent on cultivar and origin

+ Commonly held in long-term storage

Evaporative cooling from the incoming air assists in cooling and humidifying the store. This kind of store can be used for holding potatoes through the winter (three to nine months) provided they have been cured and treated with sprout suppressant. Onions and garlic can also be stored using the same techniques but with lower humidities. Garlic in California can be held for three to four months. In onions there are great differences between varieties and production locations. Both crops will need to have been dried and cured in the field. Sweet potatoes need to be cured at 28 to 30°C for a few days. Subsequently they can be stored for up to six months. Cabbages, carrots, pumpkins, apples, pears and lemons have all been successfully stored using this technique.

In Syria, unirrigated apples are stored in caves for nearly 10 months. Yam barns are common in West Africa where individual yams are tied to 2-m high vertical posts under a palm-thatch roof. In other parts of the world yams are stacked vertically on a raised platform under a straw roof. These structures are very effective, as the tubers receive adequate ventilation and are protected from both termite attack and flooding.

The extension officer can improve on-farm storage practices by training growers in correct techniques and by himself carrying out comparisons between the recommended practices and those that the farmers normally undertake. At an open day the difference between the two batches of crops can make a vivid demonstration of the validity of his recommendations. Photographs should be taken to be able to demonstrate the differences in future years.

Transport

Most growers who do not sell to traders will transport produce to market in hired lorries or pick-ups. The grower will be given a fixed price for individual use of the transport or the lorry owner will charge by the canon. In both systems there are possibly two major inefficiencies.

Firstly, if the lorry is not fully loaded the unit costs are higher. Cost savings can be achieved by improving the process of assembling produce and encouraging the sharing of transport. The extension officer may, for example, encourage growers to assemble produce on a particular day of the week at specified collection points. Larger volumes of produce can bring cost savings through both economies of scale and by attracting a number of lorry owners to encourage price competition between them.

Secondly, when transport is costed by the canon, transporters will generally overload lorries in order to maximize their income. By assembling produce at one point so as to guarantee full loads, a fixed price for the lorry can be negotiated and the growers can themselves ensure that the transport is not overloaded.

Generally the larger the individual load, i.e. the larger the truck used, the cheaper the unit/cost of transport. In Pakistan, for example, an 8-tonne lorry travelling from northern Punjab to Karachi costs 4 000 rupees, i.e. 500 rupees per tonne of produce. The recently introduced 20-tonne articulated lorries cost 7 000 rupees for the same trip, i.e. 350 rupees per tonne. Longer wheel-based vehicles reduce vibration and consequently crop damage.

Farmers who make investments in transport mainly buy small trucks. The unit costs of transport are therefore generally higher than when hiring space in a larger lorry. Pick-ups do, however, offer farmers the advantages of:

- transporting produce immediately after harvest into the market;
- supplying transport services to neighbouring growers;
- taking produce for sale at farmers' markets;
- making direct sales to retailers and catering clients.

MARKETS AND SELLING

Establishing Collection Centres and Assembly Markets

Collection centres enable produce to be assembled in volume which, in turn, attracts buyers and creates competition between buyers. Better prices are realized and economies can be achieved in transport.

In a situation where some ad hoc shipments of produce are being made into a distant market, the extension officer should consider the possibility of establishing a collection centre or assembly market. This would be achieved by:

- identifying a suitable market location, i.e. accessible to producers and roads;
- agreeing with local growers to synchronize harvesting and deliver their produce to the assembly points on a specific day of the week;
- informing buyers, agents, wholesalers and truckers of the time and location of the assembly market;
- encouraging growers not to undercut one another by agreeing on a minimum sale price.

Provided the assembly market is a success it will become self-sustaining, but at a later date it can be used as a springboard for group transport and marketing into the high-price, distant, urban markets.

Farmers' Markets and Village Markets

Farmers' markets in selected towns, where farmers or their associations can sell direct to retailers or individual consumers have been a Success in countries as different as Malaysia, North America and Pakistan.

In Egypt, highly successful village markets have been established which operate on a weekly basis. These enable farmers to sell directly either-to consumers or wholesaler agents from the city markets.

Establishing new markets is a valuable function for an extension officer because it provides new outlets for growers, increases the efficiency of middlemen and enables farmers to build Up an understanding of market orientated production.

Working with farmer groups. Working with farmer groups increases the effectiveness of an extension officer and provides opportunities for collective or cooperative action.

The success of both the above examples depended on the extension officer having formed a farmers' group through which he worked. The farmers had sufficient mutual trust to make the enterprises work.

Working with Agribusiness

The presence of an effective extension officer can attract businesses to invest in an area or into attempting to start marketing produce. The extension officer can carry out a vital intermediary role between the farmers and the agribusiness. He can organize farmers to coordinate production, provide training and negotiate contracts on behalf of the growers.

Cooperatives

In countries cooperatives have been formed to take on the marketing role on behalf of farmers. An effective cooperative will increase the chances of small farms remaining viable as the market becomes more developed and demanding. The manual has shown many examples of informal cooperation between growers. When considering the formation of a cooperative the extension officer should take into account:

- the importance of the producer retaining control. This is ensured through the cooperative's constitution. Either "one man one vote" or share capital linked to acreage is desirable;
- the need to employ efficient and well-motivated staff, particularly at senior management level. Many cooperatives have failed through employing the wrong staff and paying them according to government pay scales. Marketing requires a business mind. Successful cooperatives have often been established by linking the manager's salary to turnover and/or to the net sum paid to producer members;
- the value of a properly equipped cooperative. It must have the necessary facilities and equipment to carry out its objectives but without building up too high a level of overheads.

Cooperatives are born out of necessity, when growers recognize that their survival depends on the collective negotiating strength that a cooperative can provide.

Negotiating and selling. For the grower perhaps the most critical time in the production/marketing chain is the point when he negotiates a price with the buyer. The extension officer's role should be to improve the farmers' negotiating strength, training the grower in the art of selling and alerting him to dishonest practices.

Generally farmers are at a disadvantage in negotiation because of lack of price knowledge. Their strength in negotiation can be increased by:

- growing crops for which there is a strong demand;
- being aware of prevailing market prices and conditions and how these translate back into farmer prices;
- knowing the break-even cost of production and marketing.

The extension officer will be able to assist the farmer in these three areas with information as well as by negotiating on behalf of groups of growers and drawing up contracts.

Part of the technique of obtaining good prices for produce is good promotion and selling skills. The farmer in effect needs to be his own salesman. This will involve persuading people to buy. Advantages over other products need to be emphasized. Attractively presented and appetising produce is also effective in obtaining good prices.

Middlemen and traders can be dishonest. A bad dealer can cheat the farmer:

- on the weight of the crop;
- on the comparative quality of the produce;
- in calculating the selling price;
- by delaying payment.

The extension officer can reduce these kinds of problems by identifying honest traders in the first place. He or she can also test the accuracy of scales. (A subtle and effective method is to test if the scales measure the marketing advisor's own weight accurately.) Market news services and market knowledge will enable a farmer to be confident of how the quality of his produce compares with the competition. By encouraging the sharing of information about the reputation of traders, growers will rapidly learn which traders to trust.

A simple method for recording price information and reducing chances of cheating, when produce is sold on commission, is to have despatch notes printed with four copies, one to be retained by the producer, one for the haulier and two for the commission agent. All despatch notes are serially numbered. The producer checks with the commission agent daily by phone how the sales are progressing and will note on his copy of the despatch note the individual selling prices. When the whole consignment is sold the commission agent will return one copy of the despatch note with the selling prices recorded, along with the cheque.

Miscellaneous Activities

Horticultural research is notoriously weak in both the area of post-harvest handling and developing market-oriented production techniques. The extension officer may have an important role in identifying research topics. For example, he may encourage researchers to develop off-season vegetable production techniques, or to consider an improved packaging design and to research into onfarm crop storage techniques.

The extension officer should also play a role in assisting the planning of national projects to expand horticultural production. In addition, his advisory functions may include assisting in the planning of national market development programmes, i.e. sites and designs of new markets, marketing systems and improved infrastructure.

COMMON MISTAKES IN HORTICULTURAL MARKETING

This section examines some of the most common misunderstandings of horticultural marketing and the mistakes made.

Fixed-price Buying by Governments

Fruit and vegetable production can be highly risky with market prices sometimes being too low to cover costs. As a result there is often an outcry, particularly from growers, for the government to buy horticultural produce at fixed prices.

Whenever schemes like this have been introduced growers have responded by expanding production. They do this because government prices guarantee profitable, risk-free production which bears no relation to demand. Large volumes of produce are wasted. These schemes are a great drain on government resources which could be spent better elsewhere, for example by stimulating local demand or developing new markets.

Food Processing to Utilize Surpluses

When prices are forced down because of over-production, it is often recommended that a food-processing plant be established to utilize the surplus.Successful, profitable and self-sustaining food-processing industries cannot be based on the occasional supply of raw material when the fresh market is glutted. Horticultural processing requires expensive investment in machinery. It is crucial to make optimum use of the equipment and minimize idle time. Successful plants have to have a guaranteed supply of raw material and must produce products for which there is a demand. Food processors enter into contracts with growers to ensure that supply is evenly extended over the longest possible supply season.

Mechanized grading

It is often felt that mechanized grading will improve the quality of produce. Grading never improves quality, it merely separates qualities. High quality produce is mainly dependent on growing conditions, the production techniques used and post-harvest handling. Mechanized graders are very good at separating produce into different sizes. This is the least important aspect of grading. The most important is separating into different quality grades.

This can only be effectively done by the human eye. It is a common and recurring mistake that technology-particularly high technology-will in itself solve problems. Far greater advances can be made by improving management and adapting existing systems. Investment in expensive machinery may be a later requirement of an ongoing, successful, grading operation. If centralized grading is to work, then capital investment costs need to be minimized. Equipment such as grading tables can be made by local carpenters following a simple design. In the first instance the only mechanized equipment necessary is likely to be a conveyor which moves the produce past quality control staff. Expensive grading lines are normally only necessary if very large volumes of produce are being packed and graded for developed markets.

National Grading Standards for the Domestic Market

It is often recommended that horticultural marketing will be improved by the imposition of national grading standards. Although national standards can probably be justified for export, when compulsory minimum standards are introduced for the home market it will put up prices to the consumer. This would be entirely detrimental because it would lower consumption and reduce the size of the local market. Ideally policy measures should reduce prices and thereby increase consumption.

Grading is generally introduced by the industry itself when the consumer is willing pay a higher average price for the sorted product. Very often informal and flexible grading standards are used, which respond to the market requirements and state of supply and demand. Any attempts to force sorting on to the horticultural industry before then will be almost impossible to police, will waste government resources and will fail.

Generally the first growers to sort produce will get a high premium for their Class 1 and a low discount for their Class 3. As more growers take up grading, the increased supply of Class I may force down the price to the point where it no longer pays to grade. The viability of grading depends on whether sufficient consumers are willing to pay the extra price involved. In practice grading and sorting will take place throughout the marketing chain. Some grading, particularly of fruit, takes place in the field. Further grading often takes place at the wholesale and retail level so that damage caused by transport and delay can be dealt with. With re-grading, a balance has to be achieved between the higher price for more uniform quality and the increased damage from grading itself.

Storage of Produce to Exploit Price Rises

It is commonly thought that, in times of oversupply, produce can be held in storage and marketed when price rises occur. Most horticultural crops are only suitable for short-term storage, maybe only a few days. Storage is expensive and detracts from freshness and quality. In most situations, when the produce is brought out of store it has to compete with freshly arrived produce. The result is reduced prices and the farmer has to pay for the storage costs as well.

Relatively few crops are suitable for long-term storage. These crops should be put into storage directly after harvesting. If prices are low at harvest growers will react en masse to put a higher proportion of produce in storage. As a result when these crops are marketed out of storage there is fierce price competition. Storage in production areas is often not successful because the storage facilities are under-utilized for most of the year and are therefore uneconomic.

Government-run Trading Operations

Middlemen and traders are often accused of making excessive profits. Government-run horticultural trading operations are thought to result in improved grower returns and lower consumer prices. In practice most government-run enterprises marketing horticultural produce are failures and can only cover costs if they have some special monopoly, e.g. importing food products. Amongst the most common reasons for their failure are:

- excessive and unnecessary investments being made in equipment and buildings;
- management only working regular office hours and not having a profit incentive, unlike private-sector businesses;
- overstaffing and restrictive working practices;
- government trading operations not having sufficient flexibility to rapidly adjust prices. There is a technique to buying and selling perishable horticultural produce at a profit;
- lack of quality control and purchase of all produce offered by farmers;
- high wastage levels of produce.

Horticultural marketing is a highly competitive business requiring strong entrepreneurial and trading skills. Decisions have to be made rapidly and long hours worked, and civil services are not set up to operate in this way.

In export marketing, however, there are many examples of governments successfully establishing a single organization to oversee the development and promotion of horticultural exports. These organizations have the advantages of:

- preventing export prices being forced down by produce from the same country competing against itself;
- being able to afford highly skilled, modern managers capable of taking a strategic approach to marketing rather than a short-term, spot trading mentality typical of traders on local markets;
- being able to undertake centralized promotional and advertising campaigns;
- having large enough volumes of produce to negotiate favourable terms with shipping companies and with major buyers.

Generally the activities of support and control are allocated to the public sector, leaving the trading and business functions in the hands of the private sector.

Ultra-modern Post-harvest Techniques

Very often the level of post-harvest losses is claimed to be very high in developing countries. The introduction of post-harvest techniques used in highly developed societies, such as sophisticated packaging and cool chain techniques is expected to reduce crop wastage.

Most modern post-harvest techniques are also very expensive, requiring a high initial investment in imported equipment. They also require highly trained staff and managers and immediate access to spares and·skilled technicians.

For example, cardboard cartons can only be used once. They require considerable investment in manufacturing facilities and the continued importation of raw materials. Cool chains require specialized refrigerated stores close to the production areas in order to remove the crops' field heat, as well as refrigerated vehicles. Refrigerated containers are very expensive and the produce stored in this way should then be held in refrigerated counters even in the retail shop. The system is only practical and viable when an integrated chain is established, which requires a substantial coordinated investment, often from different firms, and when large volumes of produce are handled.

These technologies are very often inappropriate for developing countries because the costs are greater than the savings. They work best in countries with a highly developed infrastructure, i.e. good roads, reliable and cheap electricity supply, a highly skilled workforce and easy access to spare parts. Most importantly, the consumer must be prepared to pay a higher price for the produce. High technology like this does not work well in isolation. Newly introduced technologies should not be significantly more advanced than the general level of technology in a society. They should also be carefully costed to ensure that they do not add to the costs of marketing and distribution.

Change for the Sake of Change

A marketing adviser or an extension officer with special responsibilities for marketing may feel obliged to make changes to the existing marketing system. What is not often fully appreciated is that most marketing systems have evolved and will continue to respond to changing market requirements. Generally there are very good reasons why they function as they do. Like all systems a marketing system will be imperfect. However, if the system functions reasonably well, if there is competition and if produce is well distributed around the country then any marketing adviser should be extremely wary of trying to impose unnecessary changes, which may destroy what he is trying to improve. When change is not necessary, it is necessary not to change..

REFERENCES

FAO 1984. *Guide to establishing small packing stations for fruit and vegetables in rural areas.* FAO, Rome.

ITC. 1988. *Manual on the packaging of fresh fruits and vegetables.* International Trade Centre, UNCTAD/GATT, Geneva.

Kader, A. et al., 1985. *Post-harvest technology of horticultural crops.* Cooperative extension, University of California

McGee, J.R. and Kruse, M. 1986. "Swidden horticulture among the Lacandon Maya". University of California, Berkeley: Extension Media Center,.

Sargent, M.J. 1973. *Economics in horticulture,* Macmillan Press.

Thompson, S.I. 1977. "Women, Horticulture, and Society in Tropical America". *American Anthropologist,* N.S., 79: 908-910

Shepherd, A.W. 2000. *Understanding and using market information,* FAO Marketing Extension Guide No. 2, Rome.

Bibliography

Adams, C K, 2008. *Principles of Horticulture.* Butterworth-Heinemann.

Aiyer, RS. et al. 1978. "No-cost method for preserving fresh cassava roots". *Cassava Newsletter* 4: 8-9. Cali, Colombia: CIAT.

Arbury, Jim. et al. 1997, *The complete book of plant propagation,* Imprint: London: Reed International Books, 1997.

Bachmann, J. and R. Earles. 2000. "Postharvest Handling of Fruits and Vegetables". *ATTRA Horticulture Technical Note.* 19 pp.

Ball, Jeff. 1988. *Rodales's garden problem solver: vegetables, fruits, and herbs.*Emmaus, PA: Rodale Press.

Barrow, Ron. 1994. *A student handbook for the study of plant propagation - for the Advanced Certificate in Horticulture.* Published Kew, Vic. Adland Horticultural Consultants.

Berbiglia, Julie. 1996. *The Lazy Gardener'sGuide to Organic Gardening.* Scarritt-Bennett.

Bir, Richard E., 1992. *Growing and Propagating Showy and Native Woody Plants.* University of North Carolina Press.

Bird, R., 1993. *The propagation of hardy perennials,* B.T.Batsford, London.

Bradley, Fern marshall and Barbara W.Ells, editors. 1992. *Rodale's all-new encyclopidia of organic gardening: the indispensable resource for every gardener.* Emmaus, PA: Rodale Press,

Brickell, Christopher 999. *Plant propagation,* The Royal Horticultural Society. 1999.

Bryant, Geoff. 2002. *Plant propagation A-Z : growing plants for free,* Lothian.

Bubel, Nancy. 1988. *The New Seed-Starters Handbook.* Rodale Press, Inc., Emmaus, Pennsylvania.

Budelman, A. & Defoer, T. 2000. *Managing soil fertility in the tropics - PLAR and resource flow analysis in practice. Case studies from Benin, Ethiopia, Kenya, Mali and Tanzania.* Amsterdam, The Netherlands, Royal Tropical Institute.

Butt, J. 1993. *Bedding Plant Production Workshop Manual.* Crop Diversification Centre South Publication

Clarke, Graham & Toogood, Alan, 1998. *The Complete Book of Plant Propagation.* Ward Lock Limited, Wellington House, 125 Strand, London.

Crockett, James Underwood. 1972. *Vegetables and Fruits.* New York: Time Life Books.

David, Tim D, Haissig, Bruce E, Sankhla, Narendra. 1998. *Adventitious root formation in cuttings.* Published Portland, Or. : Dioscorides Press.

Denckla, Tanya. 1994. *The Organic Gardener's Home Reference: a plant-by-plant guide to growing fresh, healthy food.* Pownal, VT: Storey Communications.

Dirr, Michael A., 1987. *The Reference Manual of Woody Plant Propagation.* Varsity Press. PO Box 6301, Athens, Georgia 30604.

Doty, Walter L. 1990. *All about vegetables.* San Ramon, CA: Ortho Books.

Eames, Arthur Johnson and MacDaniels, Laurence H. 1947. *An Introduction to Plant Anatomy.* McGraw-Hill, New York;

Edmanson Jane. 1991. *Cheap and easy propagation.* Published Melbourne Lothian.

Eyasu, E. 1998. *Is soil fertility declining? Perspectives on environmental change in southern Ethiopia.* Managing Africa's Soils: 2. London, Drylands Programme IIED.

FAO. 1980. *Improving soil fertility through organic recycling - comparative analyses of organic manures,* by R.V. Misra & P.R. Hesse. Project field document No. 24. Rome.

Farabee, M.J. 2001. *Plants and their structure.* Estrella Mountain Community College, Phoenix, Arizona.

Fellows, P. and Hampton, A 1992. *Small-Scale Food Processing A Guide to Appropriate Equipment.* London: Intermediate Technology Publications. 158 pp.

Gardiner, A. 1998. *Modern plant propagation.* Published Port Melbourne Lothian.

Garner, R.J., 1985. *The Grafter's Handbook.* The Royal Horticultural Society.

Hart, Rhonda Massingham. 1991. *Bugs, slugs & other thugs: controlling garden pests organically.* Pownal, VT:Storey Communications.

Hartemink, A.E. 2001. *Soil fertility in the tropics with case studies on plantation crops.* Reading, UK, The University of Reading.

Hartmann, Hudson T., et.al. 1959. *Plant Propagation Principles and Practices.* Prentice-Hall, Inc., Simon & Schuster/A Viacom Company, Upper Saddle River, New Jersey.

Harvey, E. et al. 1990. *Harvesting and postharvest handling of papayas in the Caribbean.* Bridgetown, Barbados: Inter-American Institute for Cooperation on Agriculture (IICA).

Hawthorne, 2003. *Complete book of plant propagation,* Reed Books.

Hilhorst, T. & Muchena, F.M. 2000. *Nutrients on the move - Soil fertility dynamics in African farming systems.* London, International Institute for Environment and Development (IIED).

Ian McKiel, Erin McManus and Mary McDonald, students, University of Minnesota Department of Horticultural Sciences (2006). "Moving and Transplanting Trees with a Tree Spade". http://www.sustland.umn.edu/implement/treespade.htm. Retrieved on 2006-12-28.

Lawton, Barbara P.and George F.Van Patten. 1993. *Organic gardener'sbasics.* Van Patten Publishing.

Lima, Patrick. *The natural food garden: growing vegetables and fruits chemical-free.* 1992. Rocklin, CA: Prima Pub.

Lurie, S. 1998. "Postharvest heat treatments of horticultural crops". *Hort. Rev.* 22:91-121.

McGee, J.R. and Kruse, M. 1986. "Swidden horticulture among the Lacandon Maya". University of California, Berkeley: Extension Media Center,.

Mirza, M. and M. Younus. 1995. *Plant Nutrition and Fertilizer Management.* Alberta Tree Nursery and Horticulture Centre Publication.

Mohyuddin, M. 1990. *Greenhouse Vegetable Production Guide 1990-91.* Alberta Agriculture Publication. AGDEX 250/15 - 1.

Morard, P., A. Pujos, A. Bernadac, and G. Bertoni. 1996. "Effect of temporary calcium deficiency on tomato growth and mineral nutrition". *Journal of Plant Nutrition* 19(1):115-127.

NRC. 1992. *Neem: A Tree for Solving Global Problems.* Washington, D.C.: Bostid Publishing Co. 141 pp.

Shewfelt R.L. and Prussia, S.E. (eds). 1993. *Postharvest Handling A Systems Approach.* San Diego: Academic Press, 358 pp.

Thompson, A.K. 1996. *Postharvest Technology of Fruits and Vegetables.* Blackwell Science, Ltd., Oxford, UK, 410 pp.

University of Minnesota Extension Service 1999. *Planting and Transplanting Trees and Shrubs.* http://www.extension.umn.edu/distribution/horticulture/DG3825.html. Retrieved on 2006-12-28.

Wells, James S., *Plant Propagation Practices.* 1985. American Nurseryman Publishing Company, 111 North Canal Street, Chicago,Illinois 60606.

Wills, R., B. McGlasson, D. Graham, and D. Joyce. 1998. *Postharvest: An Introduction to the Physiology and Handling of Fruit, Vegetables, and Ornamentals.* CAB International, Wallingford, UK, 262 pp.

Zagory, D. 1999. "Effects of post-processing handling and packaging on microbial populations". *Postharv. Biol. Technol.* 15:313-321.

Index